RESURGENT COMMONS

MEANING SYSTEMS

SERIES EDITORS
Bruce Clarke and Henry Sussman

EDITORIAL BOARD
Victoria N. Alexander, Dactyl Foundation
  for the Arts and Humanities
Erich Hörl, Ruhr University Bochum
John H. Johnston, Emory University
Hans-Georg Moeller, University College
  Cork
John Protevi, Louisiana State University
Samuel Weber, Northwestern University

RESURGENT COMMONS

*Feminist Political Ecologies
in the European South*

MIRIAM TOLA

Fordham University Press : New York   2026

Copyright © 2026 Fordham University Press

All rights reserved. No part of this publication may be reproduced, stored in a retrieval system, or transmitted in any form or by any means—electronic, mechanical, photocopy, recording, or any other—except for brief quotations in printed reviews, without the prior permission of the publisher.

Fordham University Press has no responsibility for the persistence or accuracy of URLs for external or third-party Internet websites referred to in this publication and does not guarantee that any content on such websites is, or will remain, accurate or appropriate.

Fordham University Press also publishes its books in a variety of electronic formats. Some content that appears in print may not be available in electronic books.

Visit us online at www.fordhampress.com.

For EU safety / GPSR concerns: Mare Nostrum Group B.V., Mauritskade 21D, 1091 GC Amsterdam, The Netherlands, gpsr@mare-nostrum.co.uk

Library of Congress Cataloging-in-Publication Data available online at https://catalog.loc.gov.

Printed in the United States of America

28 27 26   5 4 3 2 1

First edition

CONTENTS

    Introduction: Resurgences.................... 1

    Part I: Uneasy Genealogies

1. Reading Across Commons Archives............. 27
2. Transatlantic Ecologies of Dispossession......... 46

    Part II: Production, Reproduction, Care

3. Engaging Potentials and Limits
   of the Marxist Common....................... 71
4. Transfeminist Commons:
   Inhabiting the Earth Otherwise 93

    Part III: Political Ecologies

5. Cosmopolitical Commoning
   in a City of Ruins 117
6. Crossing the Storm:
   Encountering Decolonial Perspectives........... 137

    Epilogue: On Instituting 159

    Acknowledgments 165
    Notes 169
    Works Cited................................. 183
    Index 205

RESURGENT COMMONS

INTRODUCTION: RESURGENCES

In *New York 2140*, Kim Stanley Robinson (2017) tells the story of a liquid city and a fiery political upheaval in the ruins of disaster capitalism. The novel describes a not-so-distant future in which the Antarctic ice sheet collapse has led to periods of rapid sea level rise known as the Pulses. In this scenario of climate extremes and socio-ecological disruption, the surge of water caused the flooding of coastal areas, wrecked the global supply chain, and unleashed a refugee crisis. Fossil fuel consumption slowed down significantly after the first Pulse but it was too late to avoid a second, even more devastating Pulse. As everything melted down in the increasing warmth and the higher seas, the impact of extreme weather intertwined with the violence of economic austerity producing further immiseration for the poor. New York City, the novel's epicenter and labyrinthine character, emerges from the storms transfigured into a Super Venice, crisscrossed by canals and pedestrian highlines. The superrich have moved to Uptown Manhattan, into skyscrapers on the dry land surrounding the Cloisters. Midtown and Downtown have become liminal zones inhabited by squatters and the dispossessed. Living in crumbling buildings and communal co-ops equipped with solar grids and rooftop gardens, some collectives start practicing ways of living that challenge an unsustainable political economy.

Robinson works in the tradition of the utopian genre, one in which utopia is not a perfect end-stage but the open-ended project of changing the present state of things. Imagining a path for creating futures from the predicaments of the present, his vision of underwater cities, precarious life, and wild collectives points to "the commons" as a project capable of creating alternatives to capitalist capture. Among the verdant vertical farms and the gleaming solar panels of the post-disaster city, hedge fund managers and urban developers are restlessly working to turn water fluctuations into opportunities for maximizing profit. But as a new financial bubble is swelling, an unlikely multiracial alliance of housing activists, hackers, scrappy orphans, and even a hedge fund manager, creates an unexpected opening that Robinson calls "the comedy of the commons." This novel is not an exploration of resilience, if the term is taken to mean the continuous adaptation to risks and disasters (Kaika 2017). Rather than just focusing on how to adapt to the changing climate, it clearly identifies the processes that produced the planetary crisis in the first place and indicates the commons as a possible alternative to it. Testifying to the infectious power of the commons in contemporary political and cultural imaginaries, *New York 2140* provides a fitting entry point for examining its uses in the present age of planetary crisis.[1]

This book focuses on the resurgence of the commons in contemporary political imagination and practice. Combining the history of the concept, the extended close reading of theoretical texts, and the analysis of current urban commoning practices, it examines how the commons has been reimagined and enacted as a form of self-governance that is neither public nor private. In doing so, *Resurgent Commons* shifts focus from the commons as site of tragedy, a trope popularized by Garrett Hardin in the 1960s, to its potential as alternative to forms of value extraction that deplete both territories and populations. Yet, rather than subscribing to visions of the commons as frictionless space of belonging, this work points to the ambiguities, tensions, and paradoxes within intellectual and activist practices with no guarantees of success (Berlant 2016). While I consider the commons as a political project cutting across heterogeneous times and spaces, a large part of my analysis centers on contemporary Italy, and Rome in particular, as a laboratory where intellectual production and contentious practices of commoning are intertwined, while also being in dialogue with transnational networks of radical thought and activism.[2] This vantage point allows me to trace the permutations of influential approaches to the commons, especially the work of scholars-activists associated with Italian autonomist thought, providing insights on their uneasy relationship with Western modernity. Further, researching situated urban commons in Rome enables me to make sense of the vital role of activist movements in complicating theories, producing concepts, and crafting alternatives to late capitalist intensification of social and ecological inequalities. Rome is a Southern European city, a messy striated space made up of historical layers and contradictions, where remnants of Italian colonial and fascist histories remain inscribed in the urban landscape. Largely marginal to the financial dynamics of capitalism, marked by the retreat of the welfare state and a high degree of economic precarity, it remains a major cultural center and a magnet for flows of global tourism. Rome presents aspects of "urban social extractivism," a new regime of accumulation driven by accelerated processes of profit-oriented urban revival (Rossi 2022), but it is also a "self-made city" (Cellamare 2014), proliferating with self-organized initiatives involving housing, culture, and urban natures. I analyze struggles for urban commoning as they unfold in this sprawling, unruly built environment, from transfeminist commons to collective efforts for repairing areas where industrial ruins and recalcitrant natures coexist to encounters with decolonial perspectives traveling from the Americas.

In Robinson's *New York 2140* the commons are not a merely human affair. The novel reminds us that humans do not simply create their own history on a *tabula rasa*. Nonhuman agencies—water, plants, animals, viruses, and bacteria—participate in this endeavor too. As "the citizen," the anonymous narrator who intervenes throughout the story, puts it, "people played with the global thermostat imagining they had godlike powers. They didn't" (2017, 239). Possibly, she continues, "the New York estuary was the prime actor in all that has been told here, or maybe it was bacterial communities, expressing themselves through their own civilizations, what we might call bodies" (ibid., 898). Robinson's gesture toward the more-than-human dimension of the commons resonates with this book's central concern with the relations between humans and other beings that have shaped, and continue to shape, the political ecologies of enclosures and commons. Simultaneously, I am interested in the stratifications of gender, race, class, and species that are challenged, or reproduced, in intellectual and activist modes of making commons.

The turn to the commons in contemporary movements for social and ecological justice in Rome and elsewhere provided the initial impetus for this project. A form of self-governance that is neither public nor private, the commons has long been associated with theological debates on property ownership and shared forms of land use in medieval Europe. The term has traditionally referred to the shared access to agricultural fields, pastures, and wastes that was central to medieval economies of subsistence. Over the last three decades, in a context of escalating socio-ecological injustices that closely resemble Robinson's novel, the commons have made a comeback. Amid the intensification of racial, gendered, and environmental violence produced by unnatural disasters, the expansion of extractive economies and the more recent rise of authoritarian variants of neoliberalism and new regimes of war, the commons, and the connected but distinct concepts of "common" and "commoning," have been taken up in a heterogeneous array of socio-ecological struggles. Moving between activism and academia, this capacious term has been used to describe movements defending biodiversity (Goldman 1998); campaigns opposing the privatization of water in Bolivia and Italy (Mattei 2011); anti-mining movements in Puerto Rico (García López et al. 2017); political initiatives laying claim to urban spaces from the United States to Turkey to Spain and Greece (Harvey 2012; Karaman 2013; Huron 2018; Asara 2020; Varvarousis 2022); efforts to repair the racial violence of plantations in the United States and the Caribbean (Roane 2018; Heynen 2021; Mullings 2021); and the creation of care infrastructures from below (Rispoli and Tola 2020; Ticktin 2021; Zechner 2021). Further, this concept has become central to knowledge-intensive communities including student movements protesting funding cuts for higher education (Roggero 2011), free software networks, and alternative digital currency projects seeking to support postcapitalist economies (Terranova and Fumagalli 2015; Massumi 2018). Many of these struggles cut across issues of economy and ecology. They evoke resistance to processes of financialization that subsume socio-ecological life (Cooper 2010) and extractive economies that intensify forms of exploitation and dispossession, targeting the life of populations and accelerating the wresting of raw materials from the earth on an untenable pace and scale (Gago and Mezzadra 2017; Svampa 2019).

The resurgence of the commons has been marked by the flourishing of scholarly work unearthing multiple valences of this concept. Two approaches to this topic have been particularly influential. Nobel laureate Elinor Ostrom (1990) has researched common-pool-resources (CPRs), from Swiss pastures to irrigation systems in the Philippines to forests in Japan. For Ostrom, CPRs are part of a multilevel system of governance which includes market and state-based solutions. Marxist approaches in political theory, instead, propose the commons as alternative to state and market governance. Scholars within the Italian autonomist Marxist tradition, including Antonio Negri and Michael Hardt, Paolo Virno, and Carlo Vercellone, identify "the common" in the singular as living labor, the human cooperative activities that have the potential to creating alternatives to the existing capitalist organization of life and labor (Hardt and Negri 2009; Vercellone et al. 2017). These orientations, and the ways in which they have been taken up by social movements, have been critical for reintroducing the commons in the contemporary political vocabulary. Despite significant divergences, however, these lines of inquiry have reimagined the commons through the distinction between active human collectives and malleable resources. Far from being an innocent description of the world, such a way of conceiving the commons unwittingly draws upon historically sedimented discourses and practices of European modernity that have treated nature as the mere backdrop for the (White) man's quest for progress. Given the persistent association between devalued ecologies and sexualized and racialized bodies that have been rendered as less-than-human, it is not entirely surprising that these theoretical strands often overlook how hierarchies of race, gender, species, and geography mattered, and continue to matter, in the making and unmaking of forms of life in common.

To be sure, Ostrom's theory of common-pool-resources complicates understandings of property rights and the notion of individuals as actors making rational choices about natural resources. Yet, some core tenets of modern liberalism remain central to her theory of the commons (Dardot and Laval 2015). Contemporary autonomist Marxist approaches, particularly Antonio Negri's and Paolo Virno's figurations of the common which this book discusses at length, pose significant challenges to modern liberal assumptions about property rights and individual choices as foundational for collective life. Still, they draw on a Marxist tradition that has been made up mainly by White men concerned with European property relations and the labor struggles of the White masculine working class (Heynen 2021). Even as they develop nuanced accounts of the common in advanced capitalism, they largely ignore how the differential exploitation of gendered and racialized bodies is connected to uneven histories of environmental destruction. Their expansive theories of the revolutionary potential of workers' cooperation emerging within and against globalized capital have yet to account for more-than-human lifeworlds that are simultaneously the condition of possibility of the human and its field of action.

This book asks: How did we come to think of the commons as resources or the product of human labor? How do particular bodies and beings come to matter in the making of the commons while others are relegated in the background? What different modalities of the commons come to light if we call into question the universalist figures of man-the-producer as the only subject of history? What happens when we reimagine the commons from the feminist perspective of socio-ecological reproduction (Collard and Dempsey 2018; Rispoli and Tola 2020) rather than privileging the sphere of production? Given the rich scholarly conversations about material agencies and the gendered and racialized politics of the human at a time of ecological unraveling, the time is ripe for addressing such questions. This book explores what happens when the politics of the commons encounters feminist, decolonial, and political ecology perspectives that unsettle the reduction of natures, including contemporary urban natures, to malleable matter and the setting for human interactions. It centers these perspectives for their insights about how gendered and racial violence has historically informed property relations, regulated access to the means of production and reproduction, and has been connected to the appropriation of nonhuman natures. This allows revisiting the commons, problematizing accounts that, albeit unwittingly, reinscribe universalist understandings of the White man as the primary force making the world.

Unlike accounts of the commons that focus on human activity, this book reframes the commons by attending to struggles around what, throughout the history of modernity, has been devalued as nature and confined to the realm of reproduction. In using the term devalued nature, here I refer to the construction of the material world as resources available to appropriation. At the same time, my use of the term indexes the naturalized bodies that have been placed at a distance from the hegemonic model of the human, thus creating hierarchies within the species and between humans and the nonhuman world. These include women associated to the sphere of biological reproduction and ideologies of heterosexual domesticity, Indigenous peoples deprived of relations to land and represented as savage (Coulthard 2014), Black and Brown bodies excluded from the field of the human and objectified as property (Spillers 1987; Hartman 1997; Weheliye 2014). The situated struggles for the commons analyzed in this book challenge hegemonic configurations of nature and the human to multiply alternative socio-ecological relations. They reconfigure nature as shifting assemblage of distinct beings and forces correlated in becoming which humans are part of. Part intellectual history, part close reading of texts, and part analysis of contemporary political projects, *Resurgent Commons* illuminates modes of commoning that trouble the binaries between social and natural processes, human and nonhuman, production and reproduction.

Early research for this book was conducted while the Anthropocene was becoming a keyword in the lexicon of the geosciences, helping to frame the many facets of the planetary crisis under one concept while also drawing attention from scholars across the humanities, the social sciences, and the arts. The environmental humanities emerged alongside the Anthropocene, often, although not always, highlighting the highly political dimensions of this concept. Although rejected as official epoch of geologic time, the Anthropocene retains a certain discursive power.[3] Scientific accounts of the Anthropocene, while complex and diverse, have foregrounded the geological agency of the human species, its power to impact and respond to planetary processes that have become more and more unpredictable. Feminist scholars have observed that, as much as the Anthropocene tells a story about extinction, it ultimately conveys the possibility of redemption, presenting *Homo sapiens* as a figure of both disruption and salvation. For Joanna Zylinska the Anthropocene "brings forth a temporarily wounded yet ultimately redeemed Man, who can conquer time and space by rising above the geological mess he has created" (2018, 15). In the same vein, Giovanna Di Chiro observes, "the Anthropocene retells the masculinist origin / self-birthing story that inevitably culminates in Man as the master creation, the Master of the Universe, and now its destroyer and, possibly, its saviour" (2017, 488).

As I note elsewhere (Tola 2016) the proponents of the "good Anthropocene" have taken this narrative of salvation to the extremes, rejecting dystopian visions of environmental catastrophe and suggesting that the new geological epoch is "ripe with human-directed opportunity" (Ellis 2012). From this perspective humans are the stewards of the earth or, in a twist that reflects the pervasiveness of a new logic of government, planet managers capable of converting scientific expertise into governance tools and technological fixes for navigating planetary uncertainty (Pellizzoni 2025). By relying on a troubling abstraction of man (Stengers 2015), the narrative of the good Anthropocene proposes technological solutions but has mostly refused to acknowledge the vastly uneven responsibilities in the making of ecological crises in which the burden of climate change, toxicity, waste, and ecological dispossession falls on disposable communities (Todd 2015; Collard et al. 2015; Pulido 2018; Ferdinand 2021; Barca 2020; Neimanis 2021). The "good Anthropocene" has conveniently glossed over what Françoise Vergès (2017) calls the "racial capitalocene," a term that connects the Western reduction of nature to "cheap" resources to the global organization of "cheap" gendered and racialized workforce. This narrative, in short, conceals the particularity of the subject who has historically benefited from modernity. Making this subject explicit, Di Chiro (2017) offers the term "white (M)Anthropocene" as a more precise qualifier for the current epoch.

As much as the Anthropocene has provoked conversations across academic disciplines and publics, it risks neutralizing political action. Treating the earth as an object of knowledge and intervention and establishing the primacy of geological timescales, the Anthropocene "seems to make the ordinary time of political action inoperative (if the problem is geological, what can the mere citizen do except trust experts?)" (Bonneuil and Fressoz 2016, 88). While "geocratic experts" are seen as responsible for developing technological solutions for global problems, the public is relegated to the role of responsible eco-citizen managing their individual footprint. As a project of planetary remaking, the Anthropocene presents itself as "general ecological interest" (Neyrat 2019, 2) thus leaving little room for political contestation and disagreement over the socio-ecological present (Swyngedouw 2011). What remains unchallenged is a vision of market forces as capable of addressing the environmental crisis through a range of financial instruments and valuation practices.

Resurgent Commons intersects these debates by developing an account of the commons as counterpoint to narratives that present a combination of technoscientific knowledge and market solutions as key for addressing the planetary predicament. It considers the universal *anthropos* of the Anthropocene as re-enactment of the dominant masculine subject that emerged out of the patriarchal, colonial, and capitalist relations of European modernity. The anthropos's quest for individual freedom and progress continues to depend on material relations predicated on the extractive nexus between colonial geospatial orders, the earth, and the gendered and racialized bodies that have been rendered not quite human (Yusoff 2019). In questioning the sense of the commons as an exclusively human affair, this book contributes to critical debates about political agency beyond the undifferentiated anthropos of the Anthropocene.

My account of the commons shift attention to entanglements of power, nature, and naturalized bodies, as well as the agencies that they express through political struggles. It attends to how collectives of humans and nonhumans create alternatives to processes of value extraction that target simultaneously people and the material world. In doing so, this study treats the commons through the lenses of diverse political ecologies and feminist environmental humanities.[4] I draw on speculative approaches to the political force of nonhumans (Braun and Whatmore 2010) as well as analyses of the ways in which gender, race, class, and nature have been historically bound up in power relations (Haraway 1989; Plumwood 1993; Mollett and Faria 2013). In recent years, other scholars have paid attention to the more-than-human commons. They have moved from the analysis of the commons as "thing" to the process of making commons, considering the social movements, technologies, organizational arrangements, and nonhuman others that partake in it (Bresnihan 2015; Gibson-Graham et al. 2016; Singh 2017; Centemeri 2018; Papadopoulos 2018). As much as this book resonates with these works, my primary contribution lies in historicizing shifting approaches to nature and the human in past and contemporary practices of "common use" while at the same time attending to the racial and gendered logics inscribed in these approaches. In tracing the political ecologies of the commons, I examine their complex, contradictory relations with formations of European modernity that have emerged from histories of colonization and enslavement and have been marked by racial-sexual violence and environmental destruction as conditions for capitalist accumulation. The engagement with varied and often uneasy genealogies of the commons allows me to trouble universal claims to the commons and recognize that "enclosure is not lived the same by everyone and that claims to the commons are multiple" (Eidelman and Safransky 2020, 802). It is on this basis that I then explore contemporary political projects in Rome's urban settings that infuse feminist, anti-racist, and more-than-human insights into everyday practices of commoning (Clement et al. 2019). Thus, my work intersects the rich scholarship in urban studies focusing on commons that arise in "saturated spaces" (Huron 2018), densely populated with humans and nonhumans, financial and cultural flows, public infrastructures, private profits, and conflicts over ways of making place and inhabiting the city and the earth.[5]

Resurgent Commons works across time spans and geographical spaces that might not otherwise be organized in one study. This allows the tracing of the vicissitudes, ambiguities, and paradoxes of the commons, while also considering the tensions between divergent modes of making commons. Thus, this book explores the relationship between the commons and the project of European colonial modernity with its attendant hierarchies of humanness made through relations with nature and racialized bodies rendered appropriable. Far from conceptualizing the commons simply through an oppositional relationship with the enclosures accompanying the rise of capitalist modernity, this book analyzes key authors and texts to reveal how modern assumptions about the human and nature underpin influential approaches to the commons. My objective is to document these ambiguities while also foregrounding how the commons has been reactivated through ongoing activist practices, feminist insights, and the encounter with decolonial perspectives. The remaining part of this introduction unfolds in four parts. First, I examine recent debates on the commons, the common, and the praxis of commoning and further situate this book's place in the existing literature. Then I turn to the relationship between the commons and European modernity to introduce tensions and ambiguities that will be further developed throughout this book. Finally, I offer some notes on the interdisciplinary approaches informing this research and conclude with the chapter outline.

commons, common, commoning

Published in the journal *Science* in 1968, amid rising Cold War anxieties about overpopulation and resource scarcity, ecologist Garrett Hardin's notorious essay "Tragedy of the Commons" dismissed the commons as an utterly inefficient way of managing finite resources in a world dominated by competitive individuals. In order to situate this book into larger scholarly and activist debates, it is worth briefly discussing Hardin's influential parable and then explore the differences between Elinor Ostrom's (1990) powerful rebuttal of Hardin's argument and Marxist approaches that understand *commons*, *common*, and *commoning* in terms of social relations alternative to capitalism (Hardt and Negri 2009; Federici 2004; Dardot and Laval 2015; Vercellone et al. 2017). This book is especially interested in complicating and contributing to the latter line of inquiry. It draws on its insights about capitalism's dependence on the commons while putting pressure on the relationship between labor and nature that animates it.

A wildly popular text among neo-Malthusian environmentalists and conservative policy makers, "The Tragedy of the Commons" presents a scenario in which rational herders are free to access open pastures for grazing livestock. Each individual, Hardin writes, strives to maximize his personal gain by overloading the land with no regard for the collective good: "The rational herdsman concludes that the only sensible course for him to pursue is to add another animal to his herd. And another and another . . ." (Hardin 1968, 1244). The overcrowded pasture is gradually depleted and the users impoverished: "Therein is the tragedy. Each man is locked into a system that compels him to increase his herd without limits—in a world that is limited" (ibid., 1244). In this uncompromising assessment, the commons are doomed to fail, incapacitated by man's "natural" inclination toward prioritizing individual interests over collective ones. Hardin's proposed solution toward the protection of finite resources is a combination of enclosures and coercive measures to prevent overpopulation. Not surprisingly, this tragic version of the commons has been evoked countless times by advocates of centralized state control of natural resources as well as advocates of privatization.

Much has been done to unearth the presumptions and methodological flaws of Hardin's bleak assessment of the commons. To begin with, Hardin conflates the Roman law categories of *res nullius*, vacant resources that can be appropriated, and *res communis*, that which does not belong to anyone and is unavailable to appropriation (Vercellone et al. 2017). In Hardin's essay, the herders are stripped of cultural histories, geographical locations, and social ties (Nixon 2012). Yet, they embody a pastoral version of a historically specific figure: The *homo oeconomicus* of classic political economy, a man propelled by self-interest in a world of scarcity. Imagining herders as greed-driven subjects guided by the maximization of utility, Hardin excludes the possibility of collective cooperation to prevent overloading and depletion. However, as Chapter 1 will show, historical studies of premodern Europe have demonstrated that the use of the commons was regulated through negotiations within village communities. Cooperation within localized peasant communities, rather than competition, was at the core of historical commons.

Hardin turns the grazing commons into a metaphor for illustrating a fundamental demographic concern. As postcolonial scholar Rob Nixon (2012) has noted, "The Tragedy of the Commons" was written at a time when decolonization was spreading internationally. This, I would add, was also the time when women's movements were claiming bodily autonomy from patriarchal relations. The ghosts haunting Hardin's tragic commons are the unruly bodies of poor racialized women incapable of keeping their bodies in check in a planet confronting the "population bomb" (Ehrlich 1968). "The Tragedy of the Commons" is replete not just with explicit assumptions about selfish individuals but also implicit assumptions about the gender and race of those responsible for overpopulating the planet.

The work of political scientist Elinor Ostrom (1990), recipient of the Nobel Prize for economics in 2009, provides an effective rejoinder to Hardin's argument about the unsustainability of the commons. Drawing on game theory and the systematization of extensive empirical research, Ostrom's early work demonstrates the viability of what she defines as *common-pool resources* (CPRs), that is, nonexclusive natural goods exposed to the risk of depletion. Through the investigation of long-enduring and self-organized CPRs, Ostrom shows that the successful management of CPRs entails a complex set of culturally specific arrangements that users establish and modify collectively. The governance of the commons, she argues, depends on the crafting of institutions, that is, practical rules of collective action capable of incentivizing cooperation and solve internal conflicts. From this perspective, enduring CPRs are the result of context-specific negotiations among individuals.

Ostrom's corrective to Hardin's crude application of the model of the *homo oeconomicus* has been hugely influential. She conceives of individuals as capable of joining forces for maximizing shared interests through the appropriate contractual forms. Her analysis is invaluable for clarifying questions of sustainability and durability of the commons but remains problematic in a number of ways. Building on a pluralistic model of governance, she suggests that while the crafting of enduring CPRs may work in particular cases of resource management, market and state-based solutions may be more rational or efficient in other cases. This approach, however, is not concerned with the historical erasure of the commons. Nor it is interested in the conflicts that arise in the interactions among states, markets, and commons. In other words, Ostrom's work glosses over relations of power, property, and exploitation.

Given Ostrom's articulation of CPRs as a participative governance option compatible with state and market models, her reception of the Nobel Prize in Economic Sciences in 2009 has given rise to different interpretations. Some read it as the official recognition of the viability of the commons, particularly at a time of crisis of neoliberal rationality after the global financial crisis of late 2008. Others see it as the normalization of the commons within neoliberal regimes as well as with the capitalist reappropriation of social cooperation through the so-called "sharing economy" that commodifies forms of co-living and coworking linked to digital technologies and corporate platforms (Enright and Rossi 2018). While her work on the institutions of the commons remains relevant, it is worth recalling that, as Brancaccio and Vercellone observe (2019, 707), Ostrom's main concern is "to recognize the plurality of the forms of governance and ownership, and facilitate its complementary role in the context of capitalism."

In *Capital, Volume 1* (1976) Karl Marx inaugurates a rather different approach to the commons. He famously argues that the enclosure of common land in sixteenth-century England was a key moment in the process of the primitive accumulation that prepared the ground for the transition from feudalism to capitalism. The historical process that forced peasants out of communal landholdings and into dependency on wage labor was a violent one. In Europe, it included land expropriation and the expulsion of peasants, the conversion of a variety of property arrangements into exclusive private property rights, and the commodification of labor power. In the European colonies, it comprised the devastating theft of Indigenous land and the development of massive extractive operations. For Marx, primitive accumulation was the brutal, although necessary, precondition to capitalist development. It established the primacy of private property but also created the working class, which he saw as the historical agent in the transition to socialism.

A range of thinkers have reworked Marx's linear vision of capitalist development. Focusing on the shifting power relations between the common, states, and markets, they propose the commons as an alternative to the twinned processes of capitalist dispossession and Western modernization. *Resurgent Commons* builds, and simultaneously complicates, this body of work by bringing to light some paradoxes at the core of Marxist engagements with the commons. Unlike Marx, contemporary theorists understand the "enclosures" not as limited to the early stage of capitalist development but as ongoing processes of dispossession. They argue that under neoliberal capitalism, new rounds of enclosures take place continuously and on a global scale, reconfiguring noncapitalist aspects of society into capitalist form (Mies 1989; Midnight Notes Collective 1990; Harvey 2003; Federici 2004; Linebaugh 2008). These include land-grabbing, privatization of water and land, commodification of public services, and publicly funded knowledge enclosed through patents and copyrights. In this context, the commons take on both a defensive significance as a project of resistance to neoliberal rule, and a creative significance as experimentation with noncommodified everyday-life relations of cooperation that constitute the prefigurative building blocks of a postcapitalist society (Tola and Rossi 2019). The historian Peter Linebaugh (2008) has offered the term *commoning* to index the everyday practice of making commons. Now widely adopted, this concept entails a shift from the commons as assets to be managed to the "life activity through which commonwealth is reproduced, extended and comes to serve as the basis for a new cycle of commons (re)production" (De Angelis 2017, 201). Commoning is conceived as an experimental process of struggle, one that creates new social organizations at the same time it resists the capitalist dispossession of means of subsistence, territories, and knowledges.

Scholars associated with the tradition of Italian autonomist Marxism, including Paolo Virno, Michael Hardt, Antonio Negri, Sandro Mezzadra, and Carlo Vercellone prioritize what, using the singular form, they call the *common*.[6] This term draws attention to living labor, that is, the productive dimension of social cooperation, and its defining role in current regimes of production that connect heterogeneous geographies and forms of precarious labor. For Hardt and Negri (2009), the common comprises the linguistic and affective relations of working subjectivities that capitalism incessantly expropriates. As Vercellone et al. (2017) put it, the common is simultaneously a mode of production and a mode of self-governance. According to these theorists, the common also comprises another dimension: The earth we share and the natural resources on which we depend. While they tend to describe the "natural common" in terms of scarcity and limits, the "social common" of ideas, codes, and images is limitless and reproducible, governed by a logic of abundance and proliferation (Hardt 2010). In acknowledging the distinction between the natural and the social as fraught with problems, Hardt and Negri (2009) gesture toward an "ecology of the common" that would result from relations of care and the mutual transformation between human and nonhuman beings.

The initial research for this book was prompted by Hardt and Negri's intriguing but underdeveloped formulation. It was undertaken in a spirit of convivial dissent with an approach to the common that celebrates the creative potential of human living labor but glosses over ecological questions and ultimately dissolves nature into social activity. As Kathi Weeks (2011) convincingly argues, the trouble with living labor is "that it is haunted by the very same essentialized conception of work and inflated notion of its meaning that should be called into question" (ibid., 15). Adding to this, this book contends that in framing living labor as the primary force transforming the world, the theorists of the common inherit one of the most distinctive modern aspects of Marx: The narrative of man capable of activating the latent potentials of nature for his own transformation. In this framework, the earth provides the condition of possibility of man but ultimately becomes his product. In this respect, the theorists of the common incorporate the modern distinction between human agents and a malleable nature. Questioning Marx's treatment of "the economic as an ontological category of the human" (Osborne 2005, 38), however, does not imply the dismissal of Marx. Nor does it imply the dismissal

Introduction: Resurgences 13

of labor as a key category for the analysis of current power formations. However, labor needs radical rethinking to better account for the geological and ecological conditions of human activity and the gendered and racial stratifications that inform living labor. For instance, an important body of scholarship has shown that the reconsideration of living labor is important to examine how capitalism extracts value from assemblages of nature and technologies, including biotechnologies, the use of microorganisms in the mining industry, and ecosystem services (Rajan 2006; Labban 2014; Johnson and Goldstein 2015). Thus, one of my guiding questions here is: How can we inherit from Marx's vital critique of capitalism without embracing the humanist teleology of labor as a transformative, and ultimately progressive force?[7] My work builds on autonomist definitions of the *common* and *commoning* as collective activity that is put to work in the circuits of capitalist accumulation and simultaneously exceeds them. However, I depart from autonomist theorists in that I shift attention from human living labor to the socio-ecological assemblages that compose the commons.

Silvia Federici and other Marxist feminists have brought a distinctive perspective to commoning, one that has created important openings to grasp the connection between the ongoing exploitation of devalued reproductive work and the appropriation of the earth. In the 1970s, Marxist feminists active in the Wages for Housework movement unveiled the "arcane of reproduction," that is, the key role of women's work in the process of capitalist accumulation (Federici 1975; Fortunati 1981; Dalla Costa 2019). They argued that under patriarchal capitalist relations, the work of caring for others, cleaning, cooking, raising children, and tending to the elderly has been strictly divided along gender lines and confined to the space of the household. Although essential for the reproduction of labor power, domestic and care work has been historically devalued, naturalized as a feminine vocation and made invisible. Early formulations of Wages for Housework have been called into question for overlooking the role of race in the capitalist organization of reproduction (Davis 1981). As a field of power, reproduction has been historically informed by racial hierarchies established through colonial plantation economies, institutionalized chattel slavery, and indentured servitude (Mullings 2021). Marxist feminism has also problematically centered women's domestic labor within heteronormative households, yet as Hil Malatino (2020) points out, its main insight "bears precisely on the simultaneous necessity and incalculability of care" (ibid., 44). Although marginalized with respect to the productive economic sphere, reproductive labor is labor without which the economy would collapse.

In conversation with materialist ecofeminists such as Maria Mies, Vandana Shiva, and Ariel Salleh, the Marxist feminist focus on social reproduction has extended from the exploitation of women's work to the appropriation of land and natural resources (Mies and Shiva 1993). If industrial capitalism has assumed women's reproductive labor as a necessary and yet external element for the creation of surplus value, simultaneously it has assumed the biosphere as an unlimited source of energy and matter to fuel a mode of production based on economic growth (Moore 2015). Integrating these aspects, Federici argues that feminist rearticulations of the commons and commoning entail a radical reorganization of social reproduction, a "re-enchantment of the world" that moves from the house to the land in a constant questioning not just of property relations but of our relations to ourselves, others, and a material world on which our bodies depend (Federici 2018).

My project draws on these insights even as it departs from the holistic views of the earth and the community which Federici articulates. This book contributes to revisiting questions of social reproduction and care in times of "co-producing injustices" (Sultana 2021) exacerbated by the current convergence between neoliberalism and far-right politics across the world.[8] It draws attention to the socio-ecological dimension of reproduction as an asymmetrical field of struggle where conflicting modes of sustaining life confront each other, differently organizing human activities in relation to living and nonliving ecologies. Such reframing of reproduction and care allows me to explore the commons as a mode of sustaining life that exists in tension with the uneven geographies of capitalist appropriation and exploitation.

COMMONING AND LIBERAL MODERNITY

In order to complicate ideas of the commons anchored on human rational choices or living labor, this book traces idiosyncratic threads through past and ongoing practices of "common use." The rich historiography of peasant customs in medieval Europe provides vivid accounts of the commons as a set of flexible practices that changed to resist feudal expropriations (E. P. Thompson 1993) but it rarely discusses how economies of subsistence were tied to patterns of personhood and socio-ecological relations. The work of scholars as diverse as Carlo Ginzburg and Caroline Bynum demonstrate that medieval and early modern everyday practices—both elite and popular—were grounded in a sense of the human as invested in intense exchanges with animated matter, elemental forces, and spiritual presences (Ginzburg 1992; Bynum 2011). Yet, the varied body of work on medieval and early modern materiality seldom touches on questions of property. This book brings into conversations strands of scholarship on medieval and early modern life that are not usually discussed together. Intertwining inquiries into theological debates on poverty, agricultural commons, and medieval materiality enables me to make legible past ecologies of the commons that may be useful for complicating contemporary assumptions about what today counts as commons and commoning. Rather than limiting the conceptual purchase of the commons to peasant economies destroyed by capitalist primitive accumulation, I am also interested in extending it to patterns of personhood and their relation to natural and social forces. My research on the commons in medieval and early modern Europe points toward practices of land use that, I suggest, are inseparable from premodern forms of embodiment and sense of the self. Timothy Reiss (2003) has offered the term *passibility* to describe premodern experiences of being *"embedded in and acted on"* one's communities of influence (ibid., 2). The physical world, the social world, and the divine named some of the communities of influence that "were what a person was." At the same time public and collective, they were "common to everyone qua human. They named existential spheres to which the person enlaced in them was in a reactive relation" (ibid., 2). Being passible meant to be neither active nor passive, but essentially interactive, capable of being affected and responding to the events occurring in the existential spheres.

Drawing on these insights, this study understands the historical commons as more than a way to organize shared access to land. Rather, it was a mode of living involving a range of powers and beings that did not entail a rigid distinction between individual persons and communities, subjects and objects, bodies and environments. I am not suggesting here that the premodern commons were a socio-ecological formation devoid of violence and social hierarchies. On the contrary, they existed in the larger context of European feudalism, one marked by political coercion, constant attempts at land expropriation, and sheer brutality. But living in the premodern commons meant living in a world of porous boundaries rather than one of autonomous subjects disarticulated from natural forces. By illuminating these aspects, this book questions current understandings of commons as the product of human activity, emerging out of collective protocols for managing CPRs or living labor's refusal of capitalist valorization. In doing so, it provides an opening for thinking differently about the socio-ecological relations that constitute the commons.

The rise of liberal modernity, the hegemonic ordering of the globe that emerged in the seventeenth century, was entwined with the affirmation of notions of property, personhood, and freedom that opposed the individual to the community. The question of protecting individual property and maximizing opportunities for accumulation was key in the rise of democratic nation-states, the expansion of the European colonial project, and the gradual shift from mercantilism to industrial capitalism. In the eighteenth century the consolidation of liberal economies and modes of government centered the European

White man as the civilizing subject of history (Lowe 2015) capable of turning the material world and its people into territories of "improvement." The liberty of modern man entailed a double, interconnected movement. First, there was the overcoming of nature, understood as resources to appropriate, what serves as the ground of history but is in itself outside of history. Second, it was made possible by distinct but connected forms of gendered and racialized labor exploitation in Europe and its colonies.

The self-owning liberal subject capable of transcending the "state of nature" occupies the center stage in modern political thought. Thomas Hobbes's "state of nature" is a domain of competition, self-interest, and enmity that precedes the formation of civil society. John Locke's version of the concept describes the fall of man from plenitude to degeneracy that can only be remediated through individual labor and the enclosure of land. If liberal politics demands the establishment of governments to represent and protect individual rights against the disorder of the state of nature, liberal economy calls for the appropriation of land through modulations of labor that included wage labor and a range of unfree, enslaved, and indentured labor.

Liberal notions of the human as possessive individual (MacPherson 1962) required not just the rendering of nature as property through labor but also specific configurations of gender, race, and sexuality that invested individuals' bodies and populations (McClintock 1995; Stoler 1995). The development of modern technologies of government targeted simultaneously "waste" land and "idle" inhabitants in the name of improvement and productivity, while a complex of legal, medical, and administrative technologies normatively distinguished between productive and unproductive individual conducts, civilized and uncivilized, normal and pathological, human and less-than-human.

Lisa Lowe (2015) shows that the late eighteenth- and early nineteenth-century definitions of liberal humanism were shaped through global relations of appropriation and dispossession. Through the entanglements with the African slave trade, the trade in Asian goods and people and settler colonialism, the bourgeois White man emerged as the proper form of the human, free of moving between the public and private spaces, the spaces of work and politics, and the protected domesticity of the heterosexual family. As liberal forms extolled the freedom of modern man, "other subjects, practices, and geographies are placed at a distance from 'the human'" (ibid., 3).

Liberal humanism, constructed through distinctions of race and gender, as well as through boundaries between humans and other beings, continue to shape persistent social hierarchies in the present. These stratifications also inform conceptions of what counts as politics and political community. The political ontology of modern liberalism requires the distinction between a social order made up of self-sufficient individuals who speak through politics and a natural order made up of passive objects that speak through science (Blaser 2013). As Marisol de la Cadena notes (2010; 2015a), modern politics is premised upon a "partition of the sensible" into humans and nature. Those who approximate the proper image of the human, the European White man, count in the sphere of politics. Conversely, the subjugated peoples considered closer to nature are left uncounted unless they disrupt the current partition of the sensible. This book contributes to investigating the exclusions produced at the global level by the ontologies and epistemologies of modern liberalism. It examines forms of commoning that provide counterpoints to modern liberal notions of property, personhood, and nature, while also maintaining attention to contradictions and paradoxes in the trajectories of the commons. As it attends to the political ecologies of the commons and focuses on the nonhuman entities and forces that participate in processes of commoning, this work is also alert to the limits of using the commons as a universalizing idiom of collective liberation.

Although the commons have European origins, Indigenous people in precolonial settings engaged in a variety of collective land-use arrangements that reflected specific ontologies involving humans and nonhumans living in relations of kinship obligations (Greer 2012; Kauanui 2015; Estes 2019). This is not to say that European commons coincided with Indigenous modes of living on land that, in spite of continuous colonial regimes of appropriation, persist in the effort of inventing futurity. Rather, it is to consider the partial connections between distinct formations that came into contact through processes of dispossession and resistance. Framing the relationship between the commons and Indigenous land use in terms of simultaneous proximity and divergence is useful for thinking about how specific histories of dispossession inform distinct political projects in present times. Scholarly work in Indigenous studies and colonial history, particularly interventions by Jodi Byrd, Glen Coulthard, J. Kēhaulani Kauanui, and Allan Greer challenge visions of the commons as a simultaneously idyllic formation that precedes *all* forms of modern dispossession and an idealized horizon for current struggles that seek to reappropriate what has been taken away. As Greer (2012; 2018) and Kauanui (2015) contend,[9] European settler colonialists brought to the New World ideas of commons and enclosures that contributed to the settler organization of space, agriculture, and livestock breeding. In the colonial context, the European commons translated into forms of dispossession of Indigenous land that often prepared the ground for privatization. Contemporary claims to a generalized reappropriation of the global commons tend to obscure these histories of dispossession and place Indigenous peoples' claims to land as secondary and particularistic with respect to the universalist project of commoning. Thus, they risk construing Indigenous struggles for sovereignty as regressive, a form of identity politics that stands in the way of more capacious projects of connecting differences through commons (Byrd and Rothberg 2011).[10]

Reading the premodern European forms of common use as distinct but connected to Indigenous modes of being on land opens up the space to make sense of the ambiguous relationship between the commons and the political ontology of modern liberalism. Considering the permutations of the commons, this book suggests that as a mode of using without appropriating, the commons has been the other within Western modernity, that which had to be erased or reshaped in the violent effort to reduce the material world into a site of value extraction. As forms of commons have persisted and traveled, undergoing processes of reinvention and resurgence in political theory and social movements, the contemporary political reimagination of the commons has incorporated modern elements. Influential iterations of the commons have integrated the investment in the dichotomy between human subjects that produce the common and natures that are constantly transformed. It is as if the commons has traversed European modernity to emerge transformed by the racialized and gendered logics that govern the forming of the human as a laboring species capable of continuously remaking the world. Highlighting the complex relations between the commons and Western modern imaginaries, this book challenges the universalist aspirations that animate current uses of the commons in political theory. It centers intellectual and activist practices that destabilize the established rubrics of "common-pool resources" and "the common" by investing in processes of reparation and care in which bodies and places exist in relations of mutual co-constitution. Thus, it contributes to creating more capacious socio-ecological commons rather than investing in the power of human cooperation alone.

UNDISCIPLINED METHODS

Over the past fifteen years or so, I have shared much time and space with activists and scholars engaged in diverse projects of commoning. From assemblies on the commons as a practice of world-making in New York City in the months before and after Occupy Wall Street, to short visits in Latin America, to the enduring relation with activist spaces and transfeminist organizing in Rome. The encounters I had in Italy, the United States, and Mexico have been central to crafting this study and its method of working across multiple temporalities and spaces for connecting threads of commoning. *Resurgent Commons* considers the temporality of the commons as multidirectional and simultaneous. It begins by exploring the nonmodern commons that European modernity attempted to displace and unfolds by considering contemporary struggles that evoke the commons as a mode of living otherwise (Povinelli 2012). The exploration of the political time of the commons as nonlinear opens up "the possibility of thinking about the historical as distinct from and other to the present *and* as a present living force" (Wiegman 2000, 824). Instead of accepting the commons as a stable category, my work adopts a genealogical method to examine aspects of the concept that were once relevant and are now ignored and with what effects. Bringing to the fore underanalyzed histories of the commons may be useful for shedding a new light on present struggles and the uncertain futures they may produce.

In *Society Must Be Defended*, Michel Foucault (2003) refers to his 1960s and 1970s historical studies of the asylum, the clinic, and the prison as works that ground the critique of disciplinary power into specific case studies, thus eschewing the generalizations proper to grand theories such as Marxism and psychoanalysis. He refers to them as forms of local critique that were made possible by "the insurrection of subjugated knowledges" (ibid., 7). This concept defines both scholarly research that has been marginalized and the disqualified knowledge of the insane, the patient, the delinquent, and all those who were subjected to processes of discipline. Genealogical analysis draws its force from the encounter between subjugated knowledges and builds on them in the attempt to contribute to a historical ontology of the present. This book relies on "knowledges from below" in two main ways. First, it connects medieval archives that have rarely been studied together, that is, erudite theological debates about poverty and property with accounts of peasant practices of common use. Second, it interprets contemporary theories of the commons alongside the close analysis of activist movements that reveal the partiality of views of the commons based on distinctions between subjects and resources. Interdisciplinary in nature, *Resurgent Commons* interweaves the analysis of an array of texts, from medieval writings to public speeches, with archival research, interviews with activists, and the sustained attention to the force of nonhuman entities that participate in the making of the commons. I approach the common through what Lowe (2015, 175) calls a "past conditional temporality," a term that suggests the existence of alternative conditions of possibility for the commons that were largely undone by European modernity. Additionally, I am interested in the ways in which those conditions of possibility might be reactivated in the present to create just socio-ecological futurities.

My work combines genealogical inquiry with diverse feminist perspectives on bodies, nature, and labor to interrogate the universalizing tendencies of influential theories of the commons. For some, the commons may evoke forms of belonging that in spite of claims to difference end up privileging sameness. It might appear as a reassertion of that universalism that feminist and anticolonial struggles have been fiercely and joyfully striving to dismantle. Does the commons present a newest version of the "general interest"? Does it reintroduce from the back door the notion that the class struggle takes priority over the subversion of patriarchal and colonial relations once positioned as "derivative and secondary" (Butler 1997)? These concerns should be taken seriously. Indeed, one of my goals here is to trouble the commons, introduce divergence within conceptions of the laboring human as the primary agent of social change. But I am also deeply attracted by the affirmative potential of the commons, its force as practice for making worlds otherwise. For others, the commons are associated with the politics of "small is beautiful" and seen as insufficient for generating large-scale socio-ecological change. Yet, this book centers experimental forms of commoning that combine their grounding in specific places in Italy with a structural critique of capitalism and its entanglement with patriarchy and colonialism. My point is that situated, resurgent commons serve as infrastructures of care in neoliberal contexts, including those marked by what Alberto Toscano (2023) describes as a "fascist potential." In the Italian case, this has been evident under Giorgia Meloni's far-right government, where neoliberal policies go hand in hand with state hostility toward poor, migrant and racialized populations, gender dissidents, transfeminist and climate justice movements. In such precarious conditions, the commons bridge everyday practices of reproduction and collective action (Farvadin and Robles 2025). They strive to repair lives and places, while at the same time contributing to build alliances for bringing about broader socio-ecological transformation.

Rather than a fantasy of an alternative good life that replaces broken ones, or an expression of the possibilities of life in romanticized ruins, this book offers the commons as precarious socio-ecological assemblage organizing within and against uneven processes of capitalist ruination.

CHAPTER OUTLINE

Part I of the book, titled "Uneasy Genealogies," provides readers with a historical framing of the commons. Chapter 1 begins to complicate the notion of the commons as resources by considering medieval archives of common use. Investigating the intersections between the Franciscan doctrine of common use, peasant forms of land tenure, and medieval materiality, I demonstrate that the medieval commons consisted not just in shared resources but in specific engagements with a material world regarded as mutable. Next, the chapter discusses how, in the fourteenth century, the Roman Catholic Church countered the doctrine of common use through the assertion of man's natural right to appropriate the physical world. The chapter shows that such affirmation of human mastery over the world created propitious ground for the rise of sixteenth-century European projects of enclosures and colonial conquest.

Starting from where the previous chapter left off, Chapter 2 focuses on dominant discourses of land use that in the seventeenth century connected both sides of the Atlantic. My focus are gendered and racialized discourses, including Locke's theory of property, that defined the dispossession of "unproductive" land in Europe and the Americas in terms of "improvement." This chapter demonstrates that land improvement was entangled with the affirmation of the industrious White man as the proper form of the human, entitled to appropriate the peasant commons and Indigenous land as his right. However, the chapter also highlights the contradictory role of collective use in processes of colonial dispossession thus troubling notions of the commons as an inherently emancipatory project. Next, the chapter considers how Karl Marx provided a powerful rebuttal to European improvement discourses. Yet, a close analysis of Marx's understanding of labor and nature shows that, not unlike his modern counterparts, he continued to frame nature as a realm that the laboring human transforms to fully realize his capacities. Thus, this chapter begins laying the groundwork for engaging contemporary Marxist theories of the commons.

Building on these historical insights, the second section, titled "Production, Reproduction, Care" considers the limits and potentials of the recent turn to "the common" in Italian political theory and activism. Chapter 3 shows how Italian autonomists still rely on the modern notion of labor as collective human activity that gives shape to the world. The discussion of Paolo Virno's identification of "the common" with the linguistic capacities of *Homo sapiens* demonstrates that by relying on a universalizing notion of species, Virno overlooks the racialized and gendered implications of species discourse. The close reading of Michael Hardt's and Antonio Negri's work reveals that, although they complicate the boundary between natural commons (the earth and its ecosystems) and social commons (the products of social cooperation), they end up reaffirming human labor as the primary force making the world. Ultimately, I contend that autonomist approaches inherit Marx's modern notion of man as self-inventive being. In doing so, they unwittingly participate in the Anthropocene narrative of human exceptionalism.

Providing a counterpoint to autonomist theories, Chapter 4 explores feminist intellectual and activist practices that center socio-ecological reproduction, care, and repair from below for building commons and inhabiting the earth otherwise. It discusses how ecofeminist thinkers, including Silvia Federici and Mariarosa Dalla Costa, complicate accounts of the commons as resources or mode of production. It then zooms in on transfeminist collectives that in Italy contest patriarchal violence and adapt concepts of reproduction and care for a time of overlapping social and ecological crisis. My analysis of activist practices and documents illuminates how relations between bodies and places are central in the making of transfeminist commons that engage in forms of prefigurative politics while at the same time having contentious interactions with political and economic actors.

Extending these feminist insights, the third part of the book, titled "Political Ecologies" continues exploring activist practices through the analysis of projects and encounters that draw attention to the role of nonhuman actors, including water and land, in the making of commons that challenge dominant regimes of property and production. Chapter 5 delves into the more-than-human dimension of the commons by turning to the ongoing process of repairing a dismissed chemical-textile complex in Rome. Since the mid-2000s, activists have reclaimed from gentrification the Ex-SNIA, a large area that between 1923 and 1954 was a site of labor exploitation and toxicity and now comprises postindustrial ruins, an urban lake and myriad species forming a dense urban ecology. Drawing on interviews, observation, and research on the history of the area, the chapter considers how nonhuman actors have been acting as a powerful catalyst in this collective struggle. I contend that the Ex-SNIA provides a glimpse into a mode of socio-ecological commoning that persists in a Southern European urban context marked by the increasing precarity of labor, livelihood, and environments. In reclaiming the existence of spaces exempt from value creation and commodification, this project strives to provoke transformation in the form of the commons.

Chapter 6 further probes established notions of the commons by focusing on the encounters between the commons and decolonial perspectives. Starting with an account of the "consensual invasion" of the Zapatistas in Europe and their visit in Italy in 2021, I move back and forth from Europe to the Americas, tracing how Indigenous movements complicate understandings of what counts as politics largely animating contemporary approaches to the commons. In conversation with decolonial scholarship and activism, I revisit Indigenous struggles in Southeastern Mexico and Bolivia that have been interpreted through the commons as a category of emancipation. The chapter attends to the significance of other-than-human beings, land, animals, water, and the dead, that surfaces in communiqués, public speeches, and a range of insurgent practices of protest and autonomy. In doing so, it continues to challenge the universalist aspirations of the commons, while also providing insights for enacting alliances in the face of what the Zapatistas have called "the Storm," the mounting violence of climate, racial, and gender injustices in landscapes of varied extraction. The epilogue addresses the role of institutions and instituting processes in contemporary practices of commoning and draws together the central arguments running through the chapters. It reflects on the tension between the precarity of activist commons and their generative capacity of inhabiting the world otherwise.

PART I

UNEASY GENEALOGIES

1

READING ACROSS COMMONS ARCHIVES

Ongoing scholarly and activist conversations have shown that definitions of the commons as stocks of resources that are managed over time (Ostrom 1990) are limited in that they risk obscuring the array of social relations involving commoners. The commons, Peter Linebaugh asserts, are best defined as a human relational praxis, that is, a practice of commoning that focuses on regenerating shared resources and resists market privatization (Linebaugh 2008). Such emphasis on the social dimension of the commons presents the merit to make explicit the historical and human interactions constituting the commons as well as the potential asymmetries and power relations involved in commoning processes. However, it downplays the significance of larger ecologies and relational milieus that exceed humans as social beings. Complicating this perspective, this chapter proposes to approach the commons as manifestations of historically situated relationships between a myriad of entities, human and nonhuman. Specifically, it turns to premodern commons in medieval Europe as a more-than-human assemblage bringing together peasants, heretics, witches, land, animals, plants, spirits, and a variety of entities that European modernity has relegated to the realms of folklore and religion. This historical focus brings into relief qualities of the early commons that do not readily accord with its current association to human practices or institutional arrangements to manage resources, and that have, as a result, largely been eclipsed by contemporary understandings of commoning.

I argue that considering these overlooked aspects of premodern commons is useful for reimagining the commons in current times, when "self-devouring growth" (Livingston 2019) depletes and commodifies the living and nonliving, giving rise to a range of collateral effects at the planetary level. Delving into premodern modes of inhabiting land might unsettle current theorizing about commoning that takes as self-evident the distinctions between the categories of natural and social, human and nonhuman, and living and nonliving. However, I do not wish to celebrate the precapitalist commons as a holistic mode of life that might be revived in the present. Rather, my orientation is genealogical. Instead of assuming the making of the commons to be an inherently human affair, I suggest that we slow down and interrogate how we have come to regard it in such a way and how it could be thought otherwise.[1] Inspired by Michel Foucault's (2003) genealogical inquiry, this chapter mobilizes historical practices and discourses that have been buried, ignored, and disqualified. These discourses, according to Foucault, include erudite knowledge that has been marginalized, and popular knowledge produced by those who have been deemed incapable of adequate conceptualization, such as witches, peasants, and the insane. Taken together, these perspectives constitute a memory of combats that can be reactivated to enhance contemporary commoning struggles in the ruined landscapes of late capitalism (Tsing 2015).

The legal, theological, and political significance of conflicts around property ownership and land use in medieval and early modern Europe have been the subject of much debate in critical scholarship. Karl Marx famously described the sixteenth-century enclosures of the English commons as a historical event marking the rise of capitalism. In contrast with Marx's linear account of history, some scholars contend that the enclosures are ongoing and continuous in the dynamic of capitalist accumulation (De Angelis 2007; Federici 2004; Harvey 2003). Others have investigated the medieval tradition to reflect on the significance of dissident forms of life based on common use in the present time (Agamben 2013). Less attention, however, has been paid to collectives of beings, human and other-than-human, that constituted the fabric of premodern commons. Moving in this direction, this chapter connects debates on common use that are rarely brought together. It considers the intersections between the erudite approaches to common use disqualified as heretic by the Roman Catholic Church in the thirteenth century and the popular practices of shared land use infused by magic that were largely, although not entirely, swept away through the enclosure process. Building on Lisa Lowe's insights about reading across archives in a way that "unsettles the discretely bounded objects, methods and frameworks" (Lowe 2015, 6), I engage medieval political philosophies, histories of peasant commons, and medieval conceptions of mutable matter. Thus, I demonstrate that the medieval commons were more than a way to organize shared access to the land. They expressed a mode of dwelling into a more-than-human world that did not entail a rigid distinction between subjects and objects.

Jane Bennett, a major proponent of the materialist turn in political and social theory, has argued that things have their own tendencies and propensities: Vibrant matter impinges upon humans in a world teeming with other-than-human forces (Bennett 2010). Thus, Bennett invites us to "to re-describe human experience so as to uncover more of the activity and power of a variety of nonhuman players amidst and within us" (2013, 109). This perspective is valuable in that it reorients our attention away from individual agents to disseminated agencies that disrupt and enable our doing. However, I remain unconvinced by the effectiveness of the "re-descriptive" approach in making sense of how particular modes of existence come undone under the pressure of destructive forces.[2] Drawing on a different strand of materialism, Silvia Federici's (2004) analysis of the enclosures in medieval and early modern Europe centers power struggles that while separating peasants from land also deprived women of control over their bodies and sexualities. The gendered division of labor that followed relegated women into the realm of reproduction rendering them dependent on men within the patriarchal heterosexual household. While Federici's main concern is the devaluation of women's labor leading to a major shift in reproductive relations, she connects this process to the devaluation of nature and the demise of magic as a conception of the world that was not grounded in the separation between matter and spirit. My work is indebted to Federici's heretic reworking of Marxism even as I depart from her emphasis on re-enchantment as a strategy for reimagining modes of life in common (Federici 2018).

This chapter looks at the premodern European commons as a configuration of nonproprietary relations involving disparate beings coming together in specific times and places. It explores the European origins of the commons and how notions of common use acquired a central role for heretic movements within Christianity, opening up a profound conflict with the Catholic Church. My primary goal is to read theological disputes for common use and struggles for preserving forms of shared land tenure together, demonstrating that the medieval commons ought to be understood in light of engagements with the material world that differ profoundly from modern understandings of nature as resources available for human appropriation. As historical studies have shown, medieval understandings of nature and matter where varied and complex but they shared a concern for the material world as dynamic and mutable (Ginzburg 1992; Bynum 2011; Hoffmann 2014). A wide range of things, from plants to stones and sacred objects, were seen as animate, capable of protecting or hurting human beings and inviting them into elaborate rituals and negotiations. The boundaries between subject and objects were porous as both were capable of interacting, provoking, and transforming. In this chapter I ask: What are the implications of medieval engagements with matter for thinking about the commons? How do they complicate contemporary theorizing that takes as self-evident the boundary between active human collectives and passive resources to be held in common? I am interested in bringing to the fore underexplored aspects of the commons by examining accounts and experiences of nature and things as infused by hidden powers. In the fourteenth century, the Roman Catholic Church's glorification of private property as a natural right marked a turning point in the relationship between religious movements for common use and institutional Christianity. Struggling to monopolize the truth, hegemonic Christianity claimed that man was entitled to appropriate the physical world and identified practices of poverty as heretic (Tuck 1979). This chapter argues that this affirmation of property ownership as a primary mode of being in the world contributed to the erosion of common living and the gradual shift toward nature as an external realm to extract, measure, quantify, and make productive.

COMMON USE AS HERESY

Medieval theological debates about community and poverty inherited concepts from the Roman tradition. Roman law distinguished between *res communes*, such as the sea and the air, which belonged to nobody, and *res nullius*, which are unowned but open to appropriation. The status of the *res communes* partially overlapped with that of divine things (*res divini juris*), excluded from commerce and exchange, and public things (*res publicae*) that pertained to municipalities and the citizens as a collective. They were all, in different degrees and for different reasons, inappropriable by private citizens.[3]

The Romans also distinguished between *imperium* and *dominium*. The public law concept of *imperium* indicated jurisdiction and power to govern. *Dominium*, instead, designated the exclusive individual ownership enjoyed by Roman citizens. The two concepts reflected the coexistence between the state and private citizens in the Roman republic. With the expansion of Roman imperial power, when citizens increasingly became subject to state authority rather than participants to the life of the community, the relation between *imperium* and *dominium* became problematic (Wood 2011). For example, Cicero argued for legal measures protecting individual property against the power of the community. This position was symptomatic of the tension between the two terms; it expressed the concern about the imperial infringement of *dominium* (Wolin 2004). When the Roman Empire was replaced by the feudal patchwork of jurisdictions, such tension increased exponentially. In the feudal state, government and property became contentious questions involving lords, kings, and ecclesial authorities (Wood 2011). As we shall see, the idea of *dominium* became particularly significant in fourteenth-century theological disputes on common use.

The intertwined questions of the unity of the Christian community and the sharing of goods were central to early Christian thought since *The Acts of the Apostles*. Political theorist Sheldon Wolin notes that the Christian notion of community, one "pitched to a transcendent key" (2004, 94), diverged sharply from the Greek *polis* and entertained a complex relationship to the secular political order. Early Christians thought of themselves as part of a community with transcendental qualities, superior to earthly societies. However, they also relied on a political order that guaranteed peace when the threatening forces of paganism were pushing from the outside of what was left of the Roman Empire. St. Paul's Letter to the Romans attempted to reconcile the Christian community with the political order by arguing, "the powers that be are ordained of God. Whoever therefore resisteth the power, resisteth the ordinance of God: and they that resist shall receive to themselves damnation" (Rom. 13:1–2). Augustine, who wrote the *City of God* in the aftermath of the Sack of Rome in AD 410, further elaborated on the theme of obedience to authority. The earthly city of sinners and the heavenly city of God were distinct but intertwined, one plagued by conflict and private interests, the other expressing the common good. Preserving authority and discipline in the earthly city was instrumental for creating the peace and unity of belief that made Christian life possible. The purpose of fear, allied to the doctrine of salvation, was to break the custom of evil and advance the spread of Christian truth (Augustine 1998). The theme of the community of goods among Christians figured prominently in Augustine's vivid rendering of the two cities. Only men who aspired to perfection, he suggested, had to renounce property. Others could retain earthly goods as "managers." Proper *dominium*, however, was a divine prerogative. As historian Peter Garnsey observes, Augustine was not concerned with the question of how private property was established in the first place. He was not "in the business of questioning anyone's title to land—unless, that is, they were heretics" (Garnsey 2007, 94).

With the development of the early church, and its transformation from a persecuted sect to imperial religion, ecclesiastic authorities became responsible for the government of Christ's kingdom. They managed an expanding structure of endowment and taxation that guaranteed the consolidation of the church's power. The apostolic ideal of a life in which everything was held in common became associated with monastic coenobitism. A variety of monastic rules appeared in the fourth and fifth centuries, operating in the shadow of the institutional church. Then, in the early ninth century Benedict of Aniane (ca.

750–821), in alliance with the Carolingian court, succeeded in imposing the *Rule of Benedict*, written by his patron Benedict of Nursia, as the standard for monastic life (Agamben 2013). Through this reform, the Roman Catholic Church and the emperor were able to extend their control over a variety of monastic practices and communities in the Holy Roman Empire. Yet, after the church seemed to have gained control over coenobitic life, the tension between ecclesiastic authorities and the community of believers reached a critical point.

In the eleventh century, Catholic clerics began to accuse of heresy those dissident groups that challenged the authority of the Roman curia. The Cathars and Waldensians, movements that flourished and reached a wide following in the South of France and northern Italy, embraced apostolic life against the corruption of the clergy. The Beguines, communities of women living a life of charity and mysticism, challenged the patriarchal structure of the church by eschewing the discipline of a regular order (Cohn 1970). These movements rejected or reworked the church's precepts regarding property, sexuality, and the achievement of spiritual life. They expressed a range of counter-conducts that resisted the amalgam of pastoral power and civil government that characterized the Roman Catholic Church (Foucault 2007). The papacy attempted to channel dissent through the institutionalization of Franciscanism, the mendicant movement created by Giovanni Bernardone, a former cloth merchant from Assisi who had renounced his possessions to live in poverty as a pilgrim. In doing so, however, the papacy found itself facing a mode of life based on the desire of using rather than owning things.

FRANCISCAN POVERTY: USE WITHOUT RIGHTS

The papacy's official recognition of the mendicant orders of the Franciscans and the Dominicans was part of a broader strategy of suppressing heretic revolts. But the divergent views and practices of poverty between the two orders sparked a conflict within the church. For the Dominicans, poverty was related to a disposition of the soul more than a renunciation of material things. It was secondary to the higher end of charity. Members of the Dominican order were not allowed to own anything in private but they practiced common property. In contrast, the Franciscan Rule of 1223 stated: "Let the Friars appropriate nothing for themselves, not a house, nor a place, nor anything else" (Francis of Assisi, quoted in Mäkinen 2001, 57). Franciscan life was modeled on that of Christ and the apostles who owned nothing either individually or as an order. The order's history was shaped by the ambivalent engagement with poverty, the form of life epitomized by Francis. The friars oscillated between the commitment to poverty and the practicalities of a growing organization within the institutional church. The uncompromising position on poverty led many Spirituals, the minoritarian group within the order, to be condemned and even executed as heretics and insane. Other Franciscans attenuated the radical message on property as they went on to occupy high positions in the church's hierarchies.

The controversy between Dominicans and Franciscans raged between the late thirteenth century to the middle of the fourteenth century. The Dominican friar Thomas Aquinas played an important role in it, launching an influential critique of the Franciscan practice of apostolic poverty. Human conduct, he argued, had to be modeled on immutable principles derived from divine natural laws but it should also be guided by a range of secondary precepts based on human agreement whose modification may be useful for social life (Coleman 2011). Property ownership fell into the category of secondary arrangements. Moreover, Aquinas maintained, there are two ways of considering a material object: "One is with regard to its nature, and that does not lie within human power, but only the divine power, to which all things are obedient. The other is with regard to its use. And here man does have natural *dominium* over material things, for through his reason and will he can use material objects for his own benefit" (Aquinas, quoted in Tuck 1979, 19). Thus he established a continuity between man's use and ownership of things.

The Franciscan philosopher-theologian Duns Scotus provided a major rejoinder to Aquinas. He contended that in the natural state of humanity each one was able to use what needed without exercising *dominium* over it, that is, without excluding others from using the same thing. Common use was distinct from and even incompatible with property, including common property. Unlike common use that derived from natural law, the institution of individual or common property, although created for protecting the civil order, had no natural ground. The practice of poverty, therefore, would entail a return to the state of innocence preceding the Fall in which everything was in common.

The controversy deepened already existing divisions between the majority of Franciscans, inclined to relax the interpretation of the rule, and those who insisted on observing it. *Apologia Pauperum* (1269), a text by Bonaventura, who guided the order as minister-general between 1257 to 1279 and was close to the moderate group of the Conventuals, attempted to reconcile the Franciscan abdication of property rights with the growing prosperity of the order and the church. The document sought to define the Franciscan obligation to poverty with great precision. It distinguished four categories of relation to material things: ownership (*proprietas*), possession, usufruct, and simple use (*simplex usus*) (Mäkinen 2001). For Bonaventura, the Franciscans friars committed to absolute poverty had to practice the simple use that follows from the abdication of rights. After separating use from the right to appropriate, however, he also maintained that the church was the owner of everything that was used by the friars. In contrast, Peter Olivi, a former student of Bonaventura, reinforced the ideal of poverty by arguing that the Franciscan vow involved not only the renunciation of ownership but also the commitment to *usus pauper*, that is, practical poverty and the penury of things in the everyday life. Criticized by the Conventuals, Olivi's work became popular among Spiritual Franciscans concerned about the increasing worldly involvement of the order (Burr 2001).[4]

For a while, the order enjoyed the favor of the papacy in the dispute with the Dominicans. In the bull *Exiit Qui Seminat* of 1279, Pope Nicholas III offered a detailed commentary of the Rule of Francis and concluded that Franciscan renunciation of property was "meritorious and holy" (Mäkinen 2001, 101). But when John XXII was elected pope in 1316 the equilibrium of power shifted toward the Dominicans. The new pope swiftly moved against the Spirituals whose commitment to *usus pauper* was seen as dangerous for the stability of the order and the curia (Burr 2001). Those who refused to submit to ecclesial authority were excommunicated as heretical. Many were imprisoned and burned at the stake. The papal bull *Quia vir reprobus* (1329) marked a turning point in the controversy. It affirmed the principle of natural property by claiming that human *dominium* over earthly possessions was analogous to divine *dominium* over creation. As historian Gordon Leff notes, the document glorified property by claiming that "from the beginning of time, before the creation of Eve or the laws of the kings, property had existed as a divine dispensation" (1999, 247). The implications of this position are striking: Property is natural to man even before the foundation of society. It does not emerge from the necessity of exchanging things but is a basic fact of human life. Against the Franciscan theory of use detached from property, the papacy contended that the relationship between human beings and the material world is defined by *dominium*. As the political theorist Richard Tuck eloquently puts it, "Property had begun an expansion towards all the corners of man's moral world" (1979, 22).

In his reading of the theological dispute on property, Giorgio Agamben (2013) argues that by refusing the conflation between use and property rights, the radical Franciscans invoked the principle of *abdicatio iuris*, that is, they renounced human law and the right to ownership.[5] They proposed instead a mode of living aligned to natural law, the law of God, in which humans could use things without claiming property over them. In other words, they came close to embody what for Agamben is an exemplary living practice, "a human life entirely removed from the grasp of the law and a use of bodies and of the world that would never be substantied into an appropriation" (ibid., xiii). In spite of the ambivalences within the order, Franciscan history gestures toward a notion of use that not only rejects juridical ownership but also the ontological separation between the user and that which is used.[6] Agamben focuses on the exemplary character of the heretical Franciscans for envisioning a form of life outside the grasp of the law. While he considers these living practices in isolation, as paradigmatic expressions of spiritual commitment, I am interested in how they intersect with broader social and ecological conditions that gave rise to conflicts in medieval societies. What is the relationship between Franciscan erudite practices of poverty, peasant practices of sharing land, wastes, and water, and broader medieval engagement with the material world? An entry point for addressing this question is *The Remembrance of the Desire of a Soul*, the biography of Francis by Thomas of Celano.

Written twenty years after the death of the saint, *The Remembrance* (Armstrong et al. 2000) evokes Francis's conception of an animated physical world in which all creatures—worms, birds, and stars—express divine power. Whereas other narratives portrayed Francis's itinerant life, *The Remembrance* provided him with a garden, a familiar feature in the life of saints and monastic orders. The Franciscan garden, however, was unique. Typically, European monastic gardens were marked out from their surroundings by walls or ditches or other kinds of enclosures that signaled the separation between sacred and profane spaces. Conversely, in Francis's garden no boundary existed between weeds and useful herbs, wilderness and cultivated land, owners and trespassers. According to Lisa Kiser (2003) the openness of the Franciscan garden suggested two things. First, it referred to the Franciscan idea that all beings, animal, vegetable, or mineral, have a place in the world irrespective of their value for the human. Those who conducted a life of common use renounced their possessions because they perceived themselves as creatures of God living among other creatures of God. Second, it suggested the uneasiness with enclosures and the utilitarian understanding of land. In Kiser's words, "Thomas is taking a visible stand against the increasingly widespread practice, especially in the Italy of his own time, of privatizing land that once before had been open to common rights of use" (ibid., 237). The open plot described in *The Remembrance* countered the "gardenization" of the European landscape, that is, the fragmentation and enclosure resulting from the rise of the precapitalist market economy. In Thomas's writings, the appropriation of land and the increase of soil productivity "represented a force that Franciscanism needed to oppose" (ibid., 239). The description of Francis's garden suggests a different way of inhabiting the world, one not defined by property relations. It also helps us to broaden the scope of the analysis to consider how this particular mode of apprehending and living the landscape was not limited to the Franciscans but partially overlapped with peasant land-use practices and the precapitalist economy of subsistence. *The Remembrance* does not offer a historical account of Franciscan engagements with the material world but it provides a critical perspective on the emergent trend toward the enclosure of land in thirteenth-century Italy.

The outcome of the theological dispute on property ownership contributed to creating propitious conditions for the demise of varied practices of common use that were not limited to religious orders but also concerned peasant communities. Against the Franciscan understanding of place, land, and objects, the church put forth a formidable defense of property. Pope John XXII effectively stated that, as the creature chosen by God, man had the natural right to possess the physical world. My point here is not to revive the 1960s argument by historian Lynn White (2003) in the influential and much-debated piece "The Historical Roots of our Ecological Crisis." White has famously contended that the foundational ideas for Western technical mastery of nature came from biblical doctrine, notably the medieval interpretations of it, that destroyed premodern animism and presented nature as divine gift to human beings. He also celebrated Francis as the patron saint for environmentalism, providing a pantheist alternative to the view of nature as human dominion. The White thesis, however, has rightly been criticized on the ground that it overstated the power of Christianization in purging "animism from medieval minds and environments" (Hoffman 2014, 92). Also, in focusing on the driving force of Christianity in shaping ideas and experiences of nature, it does not account for a range of other historical processes. Part of the problem with White's thesis is that, not unlike Agamben's, it focuses too much on erudite perspectives without exploring the continuities and discontinuities between religious controversies and broader social experiences of the material world. In contrast, this chapter brings into the same frame erudite knowledge that has been disqualified and the memories of communal practices buried by the rise of private property. It maintains that the fourteenth-century papal affirmation of human mastery over the world and identification of Franciscan poverty as heresy has to be placed in a larger picture. Even as the defeat of radical Franciscans contributed to establishing a vision of the world predicated upon a possessive anthropocentrism, it acquires more sense if viewed in light of a broader process of enclosures and resistance. The theological dispute on poverty and property took place at the beginning of a long process in European history that resulted in the dissolution of the feudal order, the rise of merchant capitalism, and the colonial adventure based on mercantilism. It is no coincidence that, throughout Europe, the two centuries that followed this dispute saw the intensification of peasant struggles to assert the validity of customary practices of common use against the impingement of legal codes.

To further complicate the rendering of the Franciscan order as an exemplary form of life outside property and the law, it is worth bearing in mind that the Franciscans were active participants in the colonial enterprise. Starting in the late fifteenth century, they traveled to the Americas as missionaries, papal envoys, and even inquisitors, bringing with them the imagination of a Christian world where everything was in common. This, however, did not translate into a vision of communal life involving the Indigenous people of the Americas. On the contrary, as the historian Julia McClure (2017) aptly notes, the Franciscans "contributed to engineering the ultimate dystopia of coloniality" in the New World (ibid., 11). This is important because, even as the history of the commons can be traced back to Europe, it has unfolded through global encounters connecting the European colonial project of expansion, enslavement, and dispossession to the resistance, survival, and regeneration of other people and territories. This chapter begins to trace a genealogy of the commons in medieval Europe, focusing on theological debates on poverty and property and their relation to broader socio-ecological practices in a more-than-human world. In Chapters 2 and 6, I will consider how colonial encounters have shaped the trajectories of the commons, marking continuities as well as discontinuities that can be sensed in the contemporary circulation of this formation.

PEASANT COMMONING

The distinction between the aristocracy and the commoners was a defining feature of medieval Europe. The rural commoners were either free peasants who had to turn to their lords for protection, or serfs, subject to the laws of their lords and obliged to work for them. Behind these broad categories, however, there were various local traditions and degrees of servitude and freedom regulated by custom. Both serfs and free agricultural workers living in the manor, the administrative unit of land controlled by the lord, had access to open fields, wastes, and fisheries that provided means of subsistence. Communal access to land took various forms in the plains of Western Europe including France and Italy, the southern German lands, the midland regions of England, and large parts of Scandinavia. Histories of medieval peasantry and medieval relationship with the natural world have provided vivid accounts of everyday life in the commons, customary usages, and struggles against the enclosures. According to historian Joan Thirsk (1964), the English commons comprised four elements that did not always exist together: 1) the open fields, that is, strips of arable land and meadow accessible by peasants; 2) the same strips of land that were converted in common pastures after harvest and in fallow seasons; 3) the "commons and wastes," less fertile land including marshes, moorland, and forests that were used to graze stock and gather timber, stone, coal, and a variety of food sources; and 4) the assembly of peasants, a village meeting or other decision-making body that regulated the access to the commons and related controversies. The wastes, sources of a great range of materials, were of crucial importance to the medieval economy of subsistence. They provided wood and timber for building, fencing, and the fabrication of equipment and utensils. Wood was the main source of fuel for heating, cooking, and working in craftsmen's workshops. Often the terrain of women and children, the wastes provided food and herbs for cooking and healing. These included berries, apples, mushrooms, nuts, greens, and a variety of medicinal flowers and herbs. Birds, rabbits, and other small game could be hunted and snared. They were also used as gifts and means of connection and obligation with other commoners (Neeson 1993). In clerical writings, wastes and wetlands were often portrayed as dangerous spaces inhabited by monsters, demons, and outlaws. They constituted the porous boundaries between order and chaos (Di Palma 2014).

In spite of Garrett Hardin's (1968) notorious depiction of the commons as free-for-all resources, shared land use in medieval Europe was regulated by customary restrictions and negotiations. The limits in the usages of the commons were meant to guarantee the renovation of land and the regeneration of trees. As E. P. Thompson elucidates, "agrarian custom was never fact. It was ambience. It might be best understood with the aid of Bourdieu's concept of 'habitus'—a lived environment comprised of practices, inherited expectations, rules which both determined limits to usages and disclosed possibilities, norms and sanctions both of law and neighborhood pressures" (Thompson 1993, 102). Reciprocal obligation rather than property was the central concept of feudal custom. This particular arrangement of existence, however, was not the expression of a radical egalitarianism but of a subsistence ethic that provided a minimal assurance of stability (Scott 2012). The communal economy was, to put it in Thompson's words, "parochial and exclusive" (1993, 179), a site of particular rights delimited by the boundaries of villages and parishes that could exclude strangers. But within the parish, there was a shared sense of place and ecological imbrication of the human and the natural.

The commons existed in tense relation with the law. As James Scott points out, "the theme of common property is indissolubly linked in the little tradition to the question of local custom versus law" (2012, 51). Medieval jurisprudence acknowledged the existence of the commons from when the Charter of the Forest was issued in 1217. While the Magna Carta came to be perceived as a cornerstone for establishing individual rights against the excesses of sovereign power, in the two charters the right to using the commons coexisted with individual rights with respect to autocratic behavior (Linebaugh 2008). Taken together, the charters protected the interests of the landowners and the church, acknowledged the rising role of the urban bourgeois, and established freedom of travel for merchants, while at the same time recognizing the existence of peasant practices of shared use. In the sixteenth century, however, the two charters were split: While the Magna Carta would evolve, the Charter of the Forest was left to collect dust.

The access and usages of common land were sources of constant conflicts between peasants and lords. Together with the serfs' frequent rebellion against labor, military service, and the regime of taxation imposed by the lords, peasant struggles for the commons were an integral part of the widespread resistance against feudal powers. In the thirteenth and fourteenth centuries, a time marked by the increase of population, the development of urban centers, and the expansion of market exchanges and the trading system, the "rights" that the peasants had acquired or preserved came under increasing pressure. The rise of the monetary economy provided incentive for the manorial lords to maximize their income through the sale of timber and wood. The demand for arable land was also on the rise and caused the widespread clearing of woods. In this context, disputes over access to the commons were frequent and sometimes violent. Jean Birrell (1987) documents the struggles for the commons in the English county of Staffordshire. Manorial court records from the early fourteenth century provide ample evidence of commoners' recourse to the law to protect the access to pastures and woodlands. Such an option, however, was available only to free men and those who were better off. Commoners were more often brought to court by the lords with accusations of fence breaking, raids, and the felling of trees. Sometimes groups of men showed up in court and claimed that they had broken down the fences and cut the trees to gain access to the commons. It is likely that while many of these disputes reached the court, a much larger number did not. Moreover, the documents surviving in the archives suggest that fence breaking, raids on woods, and other forms of protest were widespread in many other counties.

The effort of European monarchies to establish a uniform administration and create a homogeneous system of land ownership and registration was countered by peasant uprisings promoting highly localized social arrangements in which formal law tended to disappear. In the sixteenth century the continuous pushback against common land use turned into a fully-fledged project of dispossession that created the conditions for the emergence of capitalist relations of production. In some of the most vivid pages of *Capital*, Marx (1976) remarks that the birth of capitalism in sixteenth-century England was "written in letters of blood and fire in the annals of mankind" (ibid., 669).

Rejecting classic political economy's idyllic narrative of individual industry as the basis of capitalism, Marx describes the sheer violence of the enclosures as integral to primitive accumulation.[7] Nothing less than ruthless terrorism tore peasants from the land and turned them into proletarians, unattached sellers of labor with no choice but entering urban industries (ibid., 875). The soil, once a means of subsistence for commoners, was incorporated into capital, ushering in the transformation to industrial agriculture.[8] Protests and rebellions opposed the restructuring of property arrangements. From the sixteenth to the nineteenth centuries, Europe was, albeit unevenly, swept by riots, petitions, and all manner of mischief and obstruction to disrupt and delay the process of privatization of land. In Germany, instances of radical religious reformation merged with the struggles of rural and urban commoners to demand the restoration of customary rights in the Peasant War of 1525. The rallying cry "omnia sunt communia" is associated with the German mass uprising, specifically with Thomas Münzer, an uncompromised radical thinker of apocalypse and revolution, who joined the revolt in 1524 and became one of its most emblematic figures. The defeat of the German peasants did not put an end to the struggles for the commons. The obstinate persistence of commoning forcefully expressed a refusal of waged labor that would become almost unthinkable in a fully developed capitalist economy.

Histories of medieval customs and agrarian practices shed light on significant aspects of the commons. While they often employ the modern language of resources, at times they also draw attention to the socio-ecological dimension of the European commons. This can be seen for example in Joan Thirsk's (1964) rich account of arable land, pastures, wastes, and the peasant assembly as the four elements composing the commons. In order to avoid replicating the reduction of the commons to resources, Peter Linebaugh introduces the term "commoning" to indicate an activity that "expresses relationships in society that are inseparable from relations to nature" (2008, 279). This is important because it highlights the risks of reifying the commons and reframes it as ongoing process. Ultimately, however, for Linebaugh "commoning is a labor process" (ibid., 45), a praxis intimately related to local ecologies, but still a human praxis. While I find reframing the commons as activity to be useful, I worry that focusing on human praxis leaves the role of beings usually placed under the modern rubric of religious "belief" or "superstition" out of the picture. Exploring medieval practices of commoning allows for a reconsideration of the variety of nonhuman entities that intervened in human affairs, including spirits, stones, trees, herbs, and holy relics that conveyed specific powers. How did these entities matter in the making of the commons?

MUTABLE MATTER

Although Francis of Assisi has been interpreted as a precursor of ecological thought, his vision of the material world as infused with divine spirit was not entirely exceptional. In medieval Christian society "there persisted a pre- and non-Christian understanding of a natural world pervaded by spirit and negotiated by magic" (Hoffman 2014, 92). What from the modern perspective appears as "nature" was a highly unstable concept in medieval Europe. Until the seventeenth century, when nature became the primary object of scientific observation, there was no such thing as external and universal nature. Medieval nature "was neither unexceptionably uniform nor homogeneous over space and time" (Daston and Park 2001, 14). In the visual and written records of the medieval and early modern periods, "nature appeared to be both everywhere and nowhere" (Hanawalt and Kiser 2008, 1). Accounts of activities related to the material world, including agriculture, animal husbandry, medicine, and divination were ubiquitous, but nature was not conceived as an autonomous ontological domain with discrete boundaries. Rather, the boundaries of medieval nature were porous, often defined in relation to the supernatural, the marvelous, and the monstrous (Bartlett 2008).

In spite of the Catholic Church's efforts to purge pagan ways of inhabiting the world and affirm a cosmic hierarchy in which access to the divine was vertically mediated by ecclesiastic powers, theories and practices that challenged the Christian paradigm of creation persisted throughout the centuries. In his landmark study of the late sixteenth-century Inquisition trials of a miller from Frioul, in Northern Italy, the historian Carlo Ginzburg (1992) traces a peasant materialism that diverged from elite views of creation in a time marked by the spread of the printing press and the Protestant Reformation.[9] Identified as heretic and incarcerated, the miller known as Menocchio expressed a cosmology in which spiritual entities emerge from matter. At the beginning, Menocchio said at his first interrogation, "all was chaos . . . and out of that bulk a mass formed—just as cheese is made out of milk and worms appeared in it, and these were the angels, and among that number of angels, there was also God, *he too having been created out of that mass at the same time*" (ibid., 53). For Ginzburg the miller's confessions conveyed an elemental approach reflecting the peasant oral tradition, a powerful materialism that "rejected divine creation, the incarnation, and redemption" (ibid., 69) and had little in common with the official Christian doctrine. The archives of the Inquisition thus reveal how the popular view of the cosmos differed from the Christian dogma.

Troubling the distinction between the medieval elite culture and popular culture, Caroline Bynum (2011) contends that in the Middle Ages every social group had an intense concern and ambivalence for what she calls "holy matter," that is, matter as a site of transformation, generation, and corruption manifesting the power of the divine.[10] While there was no one understanding of matter, both those who doubted and those who believed assumed a dynamic matter, "and matter included the human body as well as the animal body, the body of the stars, or the body of wood, ash, and bone" (ibid., 283). If since the early Middle Ages a wide range of material things including herbs, bread, stones, and cloth were understood as having powers, in the late Middle Ages, new kinds of Christian "animated materiality" appeared. These included bleeding relics, animated images, paintings, and sacramentals used in ritual blessings. A key problem of medieval Christianity, Bynum notes, "was change—not the line between person and thing, or the line between life and death, but the divide between what something is (its identity) and its inevitable progression toward corruption" (ibid., 284). The nuances of the fascinating debate on medieval materiality are beyond the scope of this chapter but what these debates make clear is that in the Middle Ages the boundaries between people and things, subjects and objects, animate and inanimate were much more porous than they were in European modernity.

To explore the implications of these medieval engagements with matter for thinking about the commons it is worth considering, for instance, how common use was affirmed through the English ritual of perambulation. Practiced since the early Middle Ages and involving games, pageants, and plays, perambulations served multiple purposes. The ritual mapped the village territory by inscribing the knowledge of the landscape in the embodied memory of the participants. The lived landscape was shaped by and simultaneously shaped peasant communities by producing a sense of belonging. Seasonal collective walking allowed for a continuous renewal of customary land usages. Children's heads were knocked on boundary markers such as a ditch or a wall so that they could experience intense physical contact with significant places and remember them in case of disputes over the right to walk upon and access land (Thompson 1993; Olwig 2016). Perambulations, occasionally criticized by clergymen for blurring the line between sacrament and superstition, comprised blessings for the crops, the use of relics and holy objects to drive away wicked spirits threatening to spoil the harvest, and the sprinkling of holy water. After the enclosures, collective walks became opportunities for reestablishing the boundaries of the commons. In addition to holy objects, the peasants carried axes and mattocks for demolishing fences that had been built without the community's permission (Thompson 1993). Part of the process of making and reclaiming commons, these rituals reflect an engagement with a material world animated by a variety of nonhuman entities. The efficacy of perambulations in reestablishing common use and peasant ties to the land depended as much on the collective action of villagers as the multitude of things that they carried with them. This shows that the commons are woven of social and material relations throughout their tissue. Rather than being limited to exemplary practices of use without property, or peasant communities managing natural resources, the making and the maintenance of medieval commons involved the interplay between a range of actors, humans, and nonhumans.

Silvia Federici (2004) has drawn explicit links between premodern peasant economies in the commons and magical conceptions of the body and the material world. In *Caliban and the Witch* she notes that peasant life in the premodern commons was enmeshed in "an animistic conception of nature that did not admit to any separation between matter and spirit" (ibid., 142). Healers, herbalists, and midwives mobilized the powers of plants, animals, and stones in order to bring forth transformation. Their ability to engage with the hidden properties of living and nonliving things was a source of subsistence and power in peasant communities. This world of socio-ecological practices, however, was disqualified as superstition and persecuted as witchcraft. The *Malleum maleficarum* (1486) portrayed the witch as a sexually rapacious, poor woman, often old and uneducated. By the late fifteenth century most demonologists agreed that witchcraft was a female crime. In 1580, the French theorist of mercantilism, absolute state sovereignty, and demonologist Jean Bodin stated that women were fifty times more likely than men to succumb to diabolic temptations. A proponent of the notion that a large population is the wealth of the nation, Bodin advocated severe measures against witches and midwives whose activities interfered with population growth through forms of birth control and abortion. Targeting women's control over sexuality and reproduction through spectacular rituals of torture and death, the witch hunt ushered in a new patriarchal order in which the appropriation of women's reproductive work became central to the development of industrial capitalism. According to Federici, the witch hunt that in Europe peaked between the sixteenth and the seventeenth centuries, was part of the wider process of primitive accumulation that separated the peasants from the land through the enclosures and the violence on the communities that relied on the commons to survive. This line of inquiry, also explored by ecofeminist writers as diverse as Maria Mies (1986) and Starhawk (1982) shows that gendered land practices infused by magic were disqualified and buried. Both women's bodies and land were turned into objects of capitalist appropriation.[11]

Along these lines, this chapter argues that the medieval commons entailed not just erudite religious practices of poverty and peasants' shared access to land. This mode of existence weaved together spiritual practices, agrarian customs, modes of land use, relations of gender and sexuality, and a range of practical ways to approach a material world endowed with powers. But the world of magic was burned at the stake, eradicated by the regimes of truth associated to the affirmation of modern science and the rise of patriarchal colonial capitalism. As the contemporary neo-pagan witch and activist Starhawk has noted, "the smoke of the burned witches still hangs in our nostrils" (Starhawk 1988, 219). It fills us and informs contemporary perspectives on the commons, thus obfuscating the memories of socio-ecological relations including not just human beings managing inert resources, but also animals, plants, and objects that created meanings for and with the people engaging with them. By reactivating the memories of medieval forms of commoning, this chapter joins ecofeminist practices of "reclaiming" (Stengers 2011). This does not entail recreating the past but becoming sensitive to what has been destroyed and bring this attentiveness to socio-ecological memories in resurgent commoning struggles. Considering medieval commoning as inseparable from the engagement with a more-than-human world stimulates us to think more deeply about the distinction between subjects and objects that seems so central in standard approaches to the commons.

CONCLUSION

In this chapter I have begun to trace the vicissitudes of the commons in the passage from the Middle Ages to early European modernity. In order to unsettle contemporary accounts of the commons as primarily the product of human activity, I have read across the archives of common use, connecting medieval theological disputes to histories of peasant commons and studies of medieval materiality. By tying together erudite accounts of common use with histories of popular engagements with a material world percolating with powers, I show that metamorphic matter, "labile, changeable and capable to act" (Bynum 2011, 283) was the stuff of the commons. It was something that commoners tried to negotiate with, unleash, and control. Medieval practitioners of common use did not see themselves as autonomous subjects disarticulated from an external nature. They lived as beings embedded in spheres of relatedness comprising the physical world, the social world, the cosmic, and the divine. These spheres, or circles, preceded and constituted them as an outside that was manifest in all aspects of being a person. Timothy Reiss (2003) proposes the term *passibility* to define this "sense of being embedded in and acted on by these circles" (ibid., 2). Making commons in premodern Europe, in other words, meant collectively inhabiting a world of porous boundaries rather than one defined by the partition between human beings and natural resources managed in common.

In Europe, the disqualification of common use, through theological disputes and the enclosures, went hand in hand with the affirmation of a different sense of being human, one marked by proprietary claims in relation to one's self, others, and a natural world reduced to a source of raw materials to appropriate and transform. This particular sense of being human developed throughout European modernity, shaped by colonial encounters from the fifteenth century onward. As we shall see in Chapter 2, the civilized White man emerged in the circuits of colonial capitalism as a hegemonic model of the human, a subject entitled to reduce other people to things deprived of personhood, claim possession over the land of those who failed to establish property, and appropriate the labor of women. The defense of individual property, competition, and colonial expansion was fully articulated in the seventeenth century by the thinkers of natural rights theory. Among them was John Locke, whose theory of property is based on the enclosure of land through individual labor. In the context of European modernity, the natural right to property became the encompassing framework for making sense of the relationship between the human and the material world, the individual self and the body (Cohen 2009). Throughout the centuries, the matrix of *dominium* morphed and extended its boundaries from matters of church doctrine to Western conceptions of subjectivity as self-ownership, expanding its reach through the enclosures of common land in Europe, the bodies of enslaved people traded across the Atlantic, and the appropriation of Indigenous land in the New World.

2

TRANSATLANTIC ECOLOGIES OF DISPOSSESSION

> The notion that land requires improvement because its inhabitants are also in need of civilizational uplift, and vice versa, is no accident of history.
> —Brenna Bhandar (2018)

In 1699, the political economist Charles Davenant looked back over the seventeenth century to assess major shifts that had transformed England. Considering the past, he noted "a great deal more barren land: of that which was cultivated, very much was capable of melioration; and there were more forest, woods, coppices, commons, and waste ground than there is now, which our wealth did enable us from time to time to inclose, cultivate, and improve" (Slack 2014, 30). Davenant was a man of numbers, part of an elite group engaged in observations, measurement, and calculations concerning the national wealth and the size of population. Like William Petty, another influential economist discussed in this chapter, he was a proponent of quantitative approaches for calculating national resources and maintaining the proper ratio between hands and land. Responding to widespread poverty and the political and economic destabilization brought about by the English Civil Wars, Petty and Davenant proposed to draw on scientific knowledge and methods to increase the productivity of nature and labor so as to improve national greatness. Improvement, initially a recurrent term in memoranda, pamphlets, and parliamentary notes concerned with agricultural techniques, crop rotations, and the draining of fens, became a national slogan and a political imperative for policymakers and intellectuals (Thirsk 1984). As a process involving social and ecological dimensions, improvement moved from the realm of agriculture to education. The term conveyed progress in increasing the wealth of the nation through the transformation of nature to be achieved over time through thrift, labor, and the application of scientific knowledge (Slack 2015). In assessing how the nation had changed, Davenant identified the enclosure of the commons as central for the improvement process. The commons, the ground of the everyday subsistence for the lower classes, came to be seen as an obstacle on the path toward England's economic development.

In Chapter 1, I showed how medieval theological debates on poverty helped lay the groundwork for modern European conceptions of human subjectivity as separate from the material world and entitled to its possession. Here, I explore how narratives about enclosures and improvement were intimately connected to a specific *political ecology of dispossession* centered on the liberal figure of the human as owner of his body and labor. The term political ecology of dispossession describes the profound shift in the conception of the human and personhood in relation to the material world that occurred in seventeenth-century Europe, when the White male body came to define the enclosed location of the autonomous self, capable of turning nature into a condition of value creation through his own labor. In the liberal tradition championed by John Locke, whose political philosophy of improvement is discussed at length in the following pages, individual appropriation was regarded as man's natural right, a measure of moral virtue, and the basis of improvement.

For Locke, as well as for Petty and Davenant, improvement extended on both sides of the Atlantic, serving as a guiding principle for policies aimed at advancing England material progress through colonial adventures that laid the foundations for an empire of land and commerce. The logic of improvement traveled through the networks connecting heterogeneous spaces of dispossession and exploitation fueling colonial capitalist accumulation. The colonization of Ireland in the mid-seventeenth century provided the improvers with a laboratory for testing discourses and techniques targeting simultaneously land and populations. But it was the English expansion into the Americas that enabled their full development. Locke articulated a vision of transatlantic improvement that linked domestic and colonial agendas, advocating for increasing land productivity in both England and the New World, where White settlers claimed a natural right to enclose and cultivate the land that Indigenous people left idle. This vision rested on a possessive relationship to land as well as a racialized division of humanity between those capable of enclosing and improving the land and those seen as passively relying on the gifts of the earth.

A range of critics have interrogated the violence and exclusions underpinning the rise of modern liberalism, showing that in early modern Europe, rights and freedom came to define the universal human through attributes that coincided with the White male subject. As Uday Mehta (1999) notes, Locke, a precursor of modern liberalism, posited freedom for all but also assumed that only some were able to fully exercise it. Native populations were seen as incapable of appropriating the earth and, by extension, participating in a political community meant to protect property formations. Unsettling the liberal narrative of rights and freedom, Lisa Lowe (2015) has investigated the connections between European liberalism, the African slave trade, settler colonialism, and China trades. She examines modern liberalism as a project that promises rights, emancipation, and wage labor, while depending on global divisions and asymmetries that stratify the human to differentially distribute such liberties. While ideas of human freedom are universalized, "the people who created the conditions of possibility for that freedom are assimilated or forgotten" (ibid., 7). Modern liberal ideas of the human, in other words, subtend racial and gendered inequalities arising from interlocking histories of colonial encounters. This chapter demonstrates that they also rely on a specific political ecology, that is, on collective relations to the material world embedded with political values, and on distinctions between human and nonhuman beings. Different practices of enclosures and improvement in Europe and the colonies contributed to shape seventeenth-century concepts of freedom, individuality and property rights. Rearranging human relations with a material world made up of living and nonliving entities, these practices influenced modern political categories.[1]

In conversation with critical scholarship of early modern liberalism, this chapter examines the gendered and racialized dimensions of improvement as integral to the transatlantic political ecology of dispossession. I argue that civilizational conceptions of improvement as a project concerning land and populations informed the affirmation of the European White man as the proper form of human entitled to appropriate land. In the opening section I trace the rise of improvement, first as an agrarian technique reshaping the English landscape in the late sixteenth and seventeenth centuries, then as a distinctive feature of the English effort to enhance the profitability of both national and colonial land. Through the reading of a range of primary texts by the English improvers, I show how gendered and racialized practices of improving land moved beyond the economic sense and solidified as a moral concept predicated upon the civilizational superiority of the English way of life. Next, I consider Locke's theory of property in the context of transatlantic practices of enclosures and improvement. Locke's improvement narrative largely conflated the commons and Indigenous land, figuring them as uncultivated wastes. However, rather than thinking about the seizure of Indigenous land in the Americas as part of a process of enclosures sweeping through Europe and the colonies, I look at colonial dispossession as a distinct process in which the commons were at times paradoxically implicated. In doing so I complicate understandings of the commons as the animating core linking the anti-capitalist, anti-colonial, and anti-slavery revolts of an early modern multiracial proletariat (Linebaugh and Rediker 2000). Rather than just pointing to the paradoxes of European liberalism, I also focus on some paradoxes of the commons, thus resisting their universalization as paradigm of human emancipation.

I conclude this chapter by engaging Karl Marx's powerful rebuttal of civilizational improvement discourses. Marx showed the extent to which, behind the forward-looking vision of improvement for some lied the violence of dispossession on a world scale. He provided a complex picture of the expansive capitalist rearrangement of collective relations to the material world that occurred through the multiplication of productive forces as well as sites of extraction, production, and exchange. This analysis of the birth of capitalism remains relevant for formulating a critique of the transatlantic ecology of dispossession that linked processes of enclosures with the colonization of Indigenous land. Yet, even Marx framed nature as the realm that humans transform to realize their capacities. Conceptualizing labor as the force mediating between society and nature and activating nature's potential, he fueled the political imagination of man as self-inventive being.

EARTH IS THE MOTHER, LABOR IS THE FATHER

Recalling the uncertainties about the spelling of the term improvement, the historian Paul Slack (2015) notes that in 1510, the citizens of Coventry debated the "enprowement ... opprowment or profit" that might result from the commons and wastes (ibid., 5). The decision that followed concerned "the improvement of commons and waste grounds." Commons and wastes figured as unsettling places in the imagination of early modern English elites. Commons, forests, and swamps were often framed as hostile landscapes, resisting notions of proper and appropriate use (Di Palma 2014). They were conflated with wasteland and wilderness, conjuring up ideas of danger, lawlessness, and a condition of endemic poverty. The enclosures, in contrast, became the path toward economic and moral reform. They signified the betterment of soil and soul, indicating the way out from the chaos of the state of nature, toward the new order of productivity and profit.

In early modern England, the increase of land productivity was figured as part of the story of creation. Walter Blith (1652), the author of *The English Emprover*, a lengthy treatise detailing schemes of agricultural improvement, argued that God made mankind "to husbandize the fruits of the earth" (ibid.,

4). In Blith's view, husbandry, the art of tilling, cultivating, and managing land, had the power to connect men to the earth, "the very wombe that bears all, and the Mother that must nourish and maintain all" (ibid., 6). Using a highly gendered metaphor dating back at least to Francis Bacon, he described improvement as the unlocking of the earth's productivity that required the coupling of feminized, receptive land and a masculinized labor, seen as the active principle for the creation of wealth. As Laura Brace (2001) points out, as a womb and nurturing mother, the earth was not regarded as completely inert but it needed (men's) labor in order to release its potential. Thus, for Blith, even the most barren land "will admit of a very large improvement." Good husbandry could bring it not just "to their natural fruitfulness," but even "to a more supernatural advance than they were known to be" (Blith 1652, 16–17). Passive ownership was not enough for converting wastes into fruitful fields. What was needed was the husbandman's calling, a spirit of innovation and enterprise capable of transforming barren land. Enclosing land for tilling and pastures meant exercising active ownership. Blith's investment in improvement was motivated by the sense of chaos caused by the English Civil War of the 1640s that turned large swaths of English lands into spoiled ground. The battles between Cromwell's army and the royalist supporters devastated towns and villages, leaving thousands of people homeless and turning fields and orchards into infertile terrain. Faced with such destruction, Blith offered a detailed list of remedies. Together with the draining of fens and the watering of dryland areas, the fencing of the commons was recommended as it "prevents depopulation and advanceth all interest" (ibid., 1). Enclosures and improvement were simultaneously an economic imperative and a way of perfecting God's gift of nature.

Although the enclosures were a staple in improvement discourses, there were some notable exceptions. The Diggers, a radical movement that strenuously rejected waged labor and the enclosures, espoused the view that large-scale, intensive cultivation of the commons was necessary to feed the population. Gerrard Winstanley, a founded and prominent member of the Digger community founded in Surrey in 1649, reclaimed communal access to the soil to cultivate it. Even as he radically departed from the improvers' claim that there could be no improvement without enclosures, Winstanley argued for a more efficient form of cultivation that would allow poor peasants access to the wastelands, using their labor to enjoy the fruits of the land without reducing it to private property. The communal organization of improvement would have restored the earth to the original condition of "common treasury" and dissolved the exploitative relations involved in wage labor. Thus, the Diggers adopted elements of improvement to articulate the vision of a commonwealth government that "governs the earth without buying and selling" (Winstanley 2009, 301).

In the late sixteenth and seventeenth centuries, the publication of husbandry manuals flourished, targeting the widening public of gentry and merchant families. Authored by men who claimed firsthand expertise in agricultural techniques, they celebrated the enclosures as leading to the maximization of land productivity and increasing profits through the cultivation of land and the breeding of livestock. This literature provided abundant descriptions of men's work, thus defining estate management as a gendered practice. Men were envisioned as specialized laborers, whose husbandry skills attested to their virtuous manhood (Munroe 2008). Profit-oriented agriculture was a masculine endeavor, distinct from earlier subsistence gardening that valued women's work in the tasks of sowing, planting, and harvesting for household consumption or small-scale trade. According to Gervase Markham, the prolific author of several agricultural guides, being a husbandman meant bringing to full fruition God's blessings, the earth, and the labor for making it fruitful. It also meant to be industrious and pursuing prosperity because, as he put it, "profit is the whole aime of our lives in this world" (quoted in Munroe 2008, 146). While economic literature at times distinguished between gardens as sites of creativity and recreation for women and sites of productivity and profit for men, it also framed female thrift as the centerpiece of English domesticity. Markham's best-known work was *The English Hous-wife*, first published in 1615. This rich collection of practical advice, culinary recipes, medical remedies, and domestic chores emphatically positioned feminine thrift within the pursuit of national prosperity.[2] He described husbandry and housewifery as complementary and provided a detailed description of men's and women's work. In many ways, Markham's celebration of agrarian domesticity ran counter to larger economic shifts that encouraged consumption. Yet, it firmly reflected a gendered division of labor within the heterosexual family that was at the core of the improvement project. In addition to the enclosure of land, improvement also required an enclosed household and the distinction between the productive labor of men creating property through cultivation and the devaluation of women's work creating the conditions for it.

The Prussian émigré Samuel Hartlib and his extended circle of correspondents were important voices in debates about improvement. Hartlib, together with William Petty, Robert Boyle, and several others, circulated a proposal for stimulating the generation of wealth and advancing improvement in the post–English Civil War context. Influenced by Francis Bacon's investment in empirical investigation and the search for hard facts, they advocated the use of scientific skills and technological tools for properly cultivating and administrating land. The cultivation of the physical environment, usually gendered as female, went hand in hand with the larger program of developing natural knowledge through observation and, most importantly, experiments and interventions that would account for what Bacon called "nature altered or wrought," that is, nature modified by man (Bacon 2002, 176).[3] The confiscation of large portions of Irish land, following the Ulster rebellion of 1641 and the reconquest by Cromwell's army, was celebrated in Hartlib's circle as a great opportunity for experimenting with improvement on what was seen as a reservoir of idle land to render productive.

Petty, who Marx defined as the father of political economy, gave an important contribution to this project by developing concepts and techniques that translated the landscape into measurable resources. Petty was an active participant in the colonization of Ireland, initially as Cromwell's army's physician-general and then as the official responsible for conducting a survey of the confiscated lands so that part of them could be allocated to Cromwell's troops who were owed wages. Through his work in Ireland, he developed political arithmetic, an approach that applied quantitative techniques to economic, political, and social problems. As Paul Slack puts it, for Petty Ireland was "a 'white paper,' an open invitation to further intellectual inquiry at least as much as an opportunity for material profit" (Slack 2015, 98). Starting in 1655, over thirteen months, Petty conducted the Down Survey, an extensive collection of empirical data about Irish territory that allowed him to draw twenty-nine maps recording topographic as well as cadastral information about the landscape. The distinction between "profitable" land, including arable meadow and pastures, and "unprofitable" land comprising mountains, woodland, and marshes, was central to this endeavor. Making it possible to quantify the monetary value associated with land, the Down Survey facilitated the transfer of wealth from Irish to English hands. Petty's reward was the acquisition of huge estates, the basis of a considerable fortune. The survey enabled the Acts of Settlement that transferred approximately 8.4 million acres of land to Protestants, reducing Irish control of land to 22 percent (McCormick 2009). It paved the way for the enclosures, the commercialization of agriculture, and ambitious public works projects that sought to remake the Irish pastoral landscape into a divided agrarian one. Thus, Ireland came to embody the improvers' vision of transforming wasteland into fully enclosed and civilized gardens.

As Brenna Bhandar (2018) argues, the discourse of improvement provided the rationale for developing new regimes of racial ownership.[4] Specifically, Petty's quantification of the value of land on the basis of agricultural output extended to the differential valuation of populations. While working on the

Down Survey, he developed tools for tracing distinctions between people capable of improving land and creating wealth and racialized others unable to do so. The productivity of English farmers occupying Irish land was opposed to the idleness of brute, lazy Irish peasants. Fusing together "the value of Irish land with the value of Irish people," Petty established an economic vision of the earth and population productivity and the techniques for increasing it (ibid, 28).

Adding to this argument, I suggest that Petty's improvement had a decidedly gendered dimension. Elaborating on a recurrent gendered metaphor of the time, Petty famously stated that "Labour is the Father and active principle of Wealth, as Lands are the Mother" (Petty 1899, 67). Together, land and labor were at the basis of the value of all things. Yet, labor was the active force, necessary for making land fruitful through the proper forms of ownership and cultivation. Not only did Petty understand English forms of labor as the active principle in the cultivation of a feminized Irish land, but he also envisioned women's reproductive capacities as key to the project of turning idle Irish into industrious English to increase their value. As Ted McCormick recalls, his proposal "to transmute the Irish into English" centered on an exchange of women (2009, 212). Petty suggested forcibly transplanting to England thousands of unmarried Irish women in order to train them in the English art of conducting the household and move to Ireland an equal number of poor English women. Once there they would "naturally" tame the Irish husbands' passions and raise English children. Sex, motherhood, and the proper conduct of the heterosexual household were important ingredients in Petty's ambitious program for governing populations and changing the proportions of unproductive Irish and productive English. By improving human nature, the project of demographic transmutation complemented the improvement of land to ensure the production of wealth within a politically stable nation. Surveying land meant laying out the basis for enclosing it and governing flows of living bodies, human and nonhuman, moving across fenced land. Petty's practices of improvement brought together operations for making land into legible property and creating populations capable of properly cultivating it. These were imbricated processes, producing social stratifications as well as transforming relations to the material world. The pursuit of improvement, a central feature of the European project of modernity, entailed a gendered and racialized political ecology, that is, a mode of land use infused by specific social hierarchies. If Petty established and experimented with methods for the differential valuation of land and humans, his contemporary John Locke provided the philosophical and political arguments for an ecology of dispossession spanning both sides of the Atlantic Ocean.

TRANSATLANTIC IMPROVEMENT

In the *Two Treatises of Government*, published in 1690, John Locke provided a striking articulation of man's rightful appropriation of uncultivated land. He figured uncultivated land, or waste, as the state of nature preceding the civilized world of enclosures and improvement. For Locke, the establishment of individual ownership, occurring by mixing labor with nature, marked the threshold between the savage and the rational, the idle and the industrious, the passivity of subsistence economies and the productive activities of agriculture and trade. He understood private property as the core of subjectivity, a defining feature of proper human relations to the self and the world. According to this forerunner of English liberalism, being fully human meant claiming dominion over one's body and labor and, through them, appropriating the earth. In elaborating the theory of property as a natural right, Locke, a philosopher and a policymaker, was informed by events taking place on both sides of the Atlantic. In England, struggles over land and labor involved propertyless peasants resisting the extinction of the commons in a context of growing commercial agriculture and rapid urbanization. These conflicts unfolded alongside the expansion of transoceanic trade, regulated by the Navigation Acts that, starting in the mid-sixteenth century, followed the logic of mercantilism and established state control over international commerce. In the colonies, from the Caribbean region to the North American southeast, the appropriation of Indigenous land and the exploitation of enslaved African bodies enabled the establishment of what Kris Manjapra has called "the plantation complex," a large-scale system of mono-cropping cultivation produced for the international market (Manjapra 2020). This vast racialized political-ecological transformation was central to the project of colonial habitation created through the dispossession of land and the exploitation of racialized humans (Ferdinand 2021). It is against the backdrop of heterogenous capitalist formations that Locke developed his argument about man's right to appropriate the earth thus providing justification to the colonial project of land occupation and of racial subjection.

Locke opens the *Two Treatises* chapter "Of Property" with the assertion that everyone is born with a "right to their Preservation . . . as Nature affords for their Subsistence" (Locke 1988, 285). He continues with a claim that, as shown in Chapter 1 of this book, was widely debated in medieval thought: God has given the earth to mankind in common. Developing the hegemonic Christian argument about man's entitlement to control the material world, Locke argues that as the chosen creature of God, man has the right to establish dominium over the earth. To be human, he contends, is to be made in the image of God as rational creatures given the abilities and the duty to subdue the earth. While God had given the bounty of nature in common to humanity, he "hath also given them reason to make use of it to the best advantage of life and convenience" (ibid., 286). The earth and all its resources were given to men for their support and comfort. Thus, nature's products, animals, and plants, belong to mankind in common and no one can claim private dominion over them. But if men are to draw the greatest benefits from the earth, "there must be of necessity a means to appropriate them" (ibid., 286) and it cannot be supposed that what was given in common was supposed to always remain as such. Locke considers labor, which is proper to each man, as the human means of appropriation. This follows from the postulate that "Every Man has a *Property* in his own *Person*. This no Body has nay right but himself. The Labour of his Body, and the Work of his Hands, we may say, are properly his" (ibid., 287–88). For Locke, the individuals who apply their labor to cultivate and improve wastelands are entitled to appropriate not just the fruits of the earth but the earth itself. Because man has property in his own body and personhood, and since labor is an expression of embodied personhood, anything that he transforms becomes personal property by natural right. When man transforms nature through labor he creates property, removing it from the common: "As much Land as a Man Tills, Plants, Improves, Cultivates, and can use the Product of, so much is his Property. He by his Labour does, as it were, inclose it from the Common" (ibid., 290). Individual labor is what creates the distinction between property and the common that exists in nature but which, in itself, is of no use.

Those who work and cultivate the earth are creating value, increasing "the common stock of mankind" (Locke 1988, 206). This is because, unlike the common, cultivated land is productive. One acre of enclosed and cultivated land, argues Locke, is ten times more productive than "an acre of land, of an equal richness, lyeing wast in common" (ibid., 294). Enclosures and improvement would not just benefit individuals but the whole nation, creating general prosperity where once there was mere subsistence. After affirming that God gave the world in common to mankind, Locke introduces an important caveat: Those entitled to appropriate the earth are the industrious and rational men, not the fancy and the quarrelsome and not those who passively waited for nature to provide subsistence. As Laura Brace points out, *The Second Treatise* provides arguments against the aristocracy, living off a wealth that was unearned, as well as against those "expecting the earth to act as a womb, able to nourish and maintain everybody from her spontaneous gifts" (Brace 2004,

28). Locke's argument is based on a distinction between men capable of actively exercising their property rights for improving the land and others who let it lie uncultivated.

There has been rich debate in political theory about Locke's use of "waste" and the distinction between industrious and idle men. The relationship between Locke's theory of property, the domestic context of agrarian capitalism, and the push toward enclosures has been given considerable attention. Peter Laslett, the editor of the critical edition of the *Two Treatises of Government*, notes that Locke often employed the language of agrarian enclosures and the term *wast* in the *Two Treatises* referred to parceling out the common land in the manorial system (Locke 1988, 288). Neal Wood has challenged the image of Locke as a theorist of mercantile or industrial capitalism and connected his theory of property to structural changes leading to the emergence of English agrarian capitalism (Wood 1984). Locke's pejorative use of wastes aligned with the rising ethos of improvement: It referred to uncultivated land to be enclosed and made productive. Complicating this perspective, other scholars have unraveled the role of America in Locke's political theory, observing how colonial policies and struggles over land in America informed his definitions of property and sovereignty. James Tully (1993) contends that Locke's goal with *Two Treatises of Government* was to reject the view of Indigenous people as self-governing nations. Barbara Arneil (1996) proposes to read this work as a justification of the English colonial project in the New World. Locke's definition of labor as a source of property, Arneil argues, was influenced by his direct involvement in colonial policymaking. As a close political advisor to the Earl of Shaftesbury, Locke was appointed secretary to the Lords Proprietors of Carolina and then commissioner for the Board of Trade and Plantations. In these capacities, he played a significant role in crafting colonial policies directed at allotting land and sending "industrious" settlers to the Province of Carolina. Locke's famous formulation that "in the beginning all the World was America" (Locke 1988, 236) was not just using the New World as metaphor for an ahistorical state of nature but was concerned with English colonial claims to America.[5] Specifically, his theory of property provided support for English colonial ownership against Indigenous claims to land based on occupation. It also sought to distinguish between the sheer violence of the Spanish conquest and the more "peaceful" colonization pursued by Englishmen through industry and the affirmation of property rights.

Bridging gaps between these perspectives, I suggest that Locke conveys a vision of improvement as transatlantic project for increasing the wealth of the nation. This endeavor involved both the enclosure of the English commons and the appropriation of land in America. As English settlers laid claim to the New World, Locke's political theory relegated Indigenous territories back in the uncivilized past, contrasting it with the English civilizing mission of improvement. In his work the state of nature resembled the wilderness of America, vacant land populated by Indigenous people who dwelled on it without possessions and without laws and government. Locke's view of improvement through labor connects the commons and Indigenous land as the ground of property creation but he also introduces important distinctions, placing them in different positions on the temporal and hierarchical line of historical progress. While America could be freely claimed, the commons in England remained so "by Compact," and could not be appropriated as their use was regulated by "the Law of the Land" (Locke 1988, 293). If the Amerindians inhabited the state of nature, commoners lived at the margins of civilization, as a relic of the past standing in the way of a future built on industry and property. Although to varying degrees, both groups were regarded as uncivilized due to their reliance on the resources of uncultivated nature, which signified a failure to deploy the full potential of human industry for creating value out of natural resources. They lived on lands left to nature, whose potential productivity needed human labor in order to be unlocked. Specifically, by labor Locke meant English commercial agriculture as opposed to peasant subsistence economies and Indigenous practices of land use. Both the commons and Indigenous lands were potentially fruitful soil that had been neglected by English peasants and the Amerindians lacking the industry for moving beyond subsistence. Locke opposed their passive attitude with the righteousness of active proprietors, those who combined their labor with land to increase its productivity. In the face of neglected wastelands, improvement could render productive that which had been left to nature. Locke associates the qualities of self-possession, industriousness, and rationality with Christian men, distinguished from the Amerindians who, in his view, lacked the intellectual means and skills to improve the land through proper cultivation. But as he was also critical of the laziness of the English poor and was deeply concerned with the question of what action could be taken for turning the propertyless into a compliant workforce.

From this perspective, the common good was best served through the proper cultivation of land, which was capable of creating not just use value but exchange value (Wood 2002). The appropriation of Indigenous land and the enclosures of the commons were key steps in the civilizing mission undertaken by Englishmen both in the national territories and in the colonies. The transformation of nature into productive land and its incorporation in the monetary economy constituted the foundation of the civil society, a proper human community constituted by free and private persons. In the Lockean society, different degrees of industry, or efficacy in appropriating nature, gave men "Possessions in different Proportions" and money "gave them the opportunity to continue and enlarge them" through commercial circuits (Locke 1988, 301). What Locke fully develops, at the junction of the enclosures of the commons and the colonial enterprise, is a "distinct private, productive, and accumulative mode of appropriation as the morally superior basis of property" in which the industrious European man is entitled to "transform inert nature into an ever-expanding domain of value" (Ince 2018, 39) at the expenses of idle commoners and Indigenous people.[6]

This framework, subtending a possessive relation between a group of individuals and the material world, and its superiority with respect to commoners and Indigenous people, provided justification to practices of dispossession aimed at turning uncultivated earth into value-generating property. The enclosures of land laying waste in common was a moral and economic imperative, a civilizing enterprise capable of transforming both the earth and the people dwelling on it. Locke's politics and philosophy of improvement relied on an ecology of dispossession that connected the Old and the New World, demanding the stripping of English peasants from their sources of livelihood and unfettered access to Indigenous land to fully reshape the earth into a productive landscape. Yet, even as improvement discourses were concerned with dismantling both the European common and Indigenous forms of land tenure, it is crucial to pay close attention to specific forms of dispossession and avoid conflating differential modes of enclosing land in early modern Europe and the Americas. Historical research shows that European settlers employed common property as a tool for seizing Indigenous land in a range of New World colonies (Greer 2012; 2018). As I show in the next section, attention to such process matters in order to avoid positing the commons simply as the opposite of the enclosures and romanticizing the commons as spaces devoid of power relations.

PARADOXES OF THE COMMONS

Locke's arguments and other improvement narratives see the enclosures as the driving force for shaping land into productive individual property. Enclosures and commons are most often presented as mutually exclusive. An understanding of the relation between commons and enclosures in oppositional terms can be detected also in contemporary scholars of the commons who consider the brutal imposition of private property as a movement connecting the Old World and the New World. Peter Linebaugh and Markus Rediker (2000) contend that the expropriation of the commons by enclosure and conquest took place on the two sides of the Atlantic, setting in motion a series of social and economic changes including the shift from economies of subsistence to commercial agriculture, the increase of wage labor, the growth of urban populations, and the expansion of global trades and markets. They describe how in the late sixteenth and early seventeenth centuries, thousands of people were removed from the English commons and redeployed to towns and sea, often serving the cause of colonial expansion. At the same time, in America, settlers cleared the ground for agricultural colonies, "building fences and hedges, the markers of enclosure and private property" (Linebaugh and Rediker 2000, 44). Discussing the traffic in people and goods linking England, the mainland colonies of North America, the Caribbean islands, and West Africa in the early modern period, Linebaugh and Rediker highlight how the infrastructures of mercantile capitalism were built through transatlantic circuits of expropriation linking the ruthless exploitation of workers and enslaved people and the dispossession of Indigenous land. This work also illuminates the resistance of English sailors, pirates, fugitive African slaves, and Indigenous Americans, forming a "many-headed hydra," a loose alliance of rebels that alarmed the ruling classes of the day. In doing so, Linebaugh and Rediker challenge the tendency within much labor historiography to focus on the White male working class by providing glimpses of a diverse anti-capitalist tradition, and modes of living without waged labor, property, racial and gender hierarchies. Linebaugh and Rediker provide valuable historical framing to the view of the enclosures as the driving, ongoing force fueling capitalist accumulation, and the commons as the animating core of movements seeking alternatives to capitalism. However, focusing on a long-term global movement toward privatization, they overlook some of the paradoxes of commoning and present it as an emancipatory relation connecting disparate groups in the shared resistance to capitalist dispossession.

Historical research has complicated this understanding of the commons as overly romanticized and highlighted a more ambiguous relationship between commons and enclosures in the early modern colonial project. Drawing on research on seventeenth-century New Spain, New France, and New England, Allan Greer (2012) argues that privatization of land was not the only, or even the main means of dispossession of Indigenous territory. Dispossession often occurred through the "interplay between enclosure and commons" (ibid., 366). While in the long run the tendency was toward enclosed private property, the colonization process also included varieties of "colonial commons." As colonizers claimed possession of lands overseas, they did so not just through fences and hedges but often through a combination of private property and collective land use. Settler property formations encroached on Indigenous land, transforming the landscape and undermining Indigenous relations to place and to the animal and spiritual entities inhabiting it.

Although the commons is an European concept, Greer, as other scholars before him (Cronon 2003), uses the term to indicate Indigenous arrangements "by which terrain and resources belonged to specific human collectivities engaged in varied combinations of agriculture, foraging, hunting, and fishing" (Greer 2012, 32). As White settlements spread in North America, Indigenous nations "engaged in an unequal struggle to preserve their indigenous commons against the aggressive expansion of the colonial commons" (ibid., 379). A particularly destructive force of dispossession was the colonial "outer commons," a large area of settler hunting, timbering, foraging, and grazing that radically transformed Indigenous landscapes. In New Spain, for example, the Spanish version of the commons was based on ranging cattle and the introduction of sheep and goats that rapidly turned the Indigenous landscape of croplands, woodlands, and irrigation into a semidesert. In New England, as Virginia DeJohn Anderson has shown (2004), settlers running short of labor used a system of private ownership but communal management of livestock. They allowed cattle, horses, and hogs to roam free rather than keeping them into enclosed fields. Free-range livestock damaged native crops and medicinal plants thus destroying the habitat of indigenous game, provoking soil erosion, and prompting the spread of Old World diseases. Domesticated animals, seen by English settlers as part of human dominion over the material world, participated in a multispecies assault on Indigenous land and Indigenous peoples' mode of relating to the nonhuman world. Livestock was mobilized within the European political ecology of dispossession and played a role in the extension of settler dominion over America.

These historical perspectives bring into focus the ambivalent relation between the common and the enclosures, beyond an oppositional framework that posits the enclosures as always erasing commons and the commons as always resisting the enclosures. They also nuance the relationship between the enclosures of peasant land in Europe and the appropriation of Indigenous land in the Americas. These processes can be read as distinct but connected forms of dispossession disrupting situated collective relations to the material world and not simply as part of the same process of capitalist expansion erasing a universal commons. Grappling with the ways in which common use was folded into the European project of appropriating Indigenous land, helps to avoid generalizations about the emancipatory potential of restoring the commons.

In making this argument, one that I will extend in Chapter 6, I am inspired by Indigenous and decolonial scholars that invite nuanced analysis of interlocked but situated experiences of dispossession and resistance. Eve Tuck and

Wayne Yang (2012) have argued against "the homogenization of various experiences of oppression" (ibid., 3) and shown that conflating capitalism, colonialism, and other oppressive formations hinders the advancement of meaningful anti-colonial and decolonial projects. Dené political theorist Glen Coulthard (2014) has called for distinguishing the analysis of capitalist relations from colonial relations in order to develop more accurate ways to understand their intersections. These decolonial insights are relevant for the purposes of this book as they challenge visions of the enclosure as a linear movement and ideas of the commons as spaces of collective freedom. The analysis of enclosures and commons, in other words, requires attention to contradictions and differences in order to understand their specificities in particular times and spaces.

Taking into account the nonequivalence between the commons and Indigenous land use, the early modern commons can be defined as a collective mode of relations to the material world that originated in Europe and traveled to the New World, becoming implicated in projects of colonial dispossession. As colonizers traveled, they brought to America weapons, agricultural technologies, viruses, livestock, plants, and also commons that combined into an assemblage of dispossession threatening Indigenous ways of inhabiting the land. While in Europe the commons had to be erased, minimized, and relegated to the past for improvement to succeed, aspects of it could be transposed into the colonial project of appropriating Indigenous land seen as unproductive. Thus, rather than understanding the commons as an inherently innocent formation and the practice unifying disparate struggles on both sides of the Atlantic, I am interested in highlighting its ambivalent relations to early modern liberalism. On the one hand, the commons had to be cast at the margins of modernity, as a mode of relating to land belonging to a past superseded by the cumulative progress spearheaded by European industry and rationality. On the other hand, elements of commoning were selectively incorporated into various colonial practices for asserting control over Indigenous land. At the same time, they persisted as a form of resistance against the affirmation of private property regimes.

The connection between the extinction of the European commons and the expansion of European colonialism persisted in the eighteenth century. Traces of it are found in English debates on the enclosures that reiterated the tropes of the commons as wild and unproductive land inhabited by a mischievous race of lazy and dangerous people, standing in the way of improvement. John Sinclair, then president of the Board of Agriculture, wrote in 1803: "Why should we not attempt a campaign also against our great domestic foe, I mean the hitherto unconquered sterility of so large a proportion of the surface of the kingdom? . . . let us not be satisfied with the liberation of Egypt, or the subjugation of Malta, but let us subdue Finchley Common" (quoted in Neeson 1996, 31). Here, the persistence of the commons is described as a hurdle hindering the project of strengthening the vital energy of the nation. As shared forms of land use stood in the way of national interest and prosperity, the enclosures were seen as a solution to both economic and moral concerns. The establishment of property rights addressed questions of labor supply and productivity but also of poverty and criminality. In England starting in the middle of the eighteenth century, Acts of Parliament sanctioned the enclosures. Despite the stubborn noncompliance of the commoners and the struggles for land and freedom in the Americas, the civilizing politics of improvement prevailed as one of the driving forces of European modernity.

MARX, LAND, AND THE LABOR OF SPECIES-BEING

As shown throughout the preceding pages, seventeenth- and eighteenth-century discourses of agricultural improvement celebrated the role of individual thrift and ingenuity for increasing the wealth and prosperity of the nation. Improvement discourses distinguished between self-possessive subjects capable of making land valuable and unfit subjects (commoners and Indigenous people) living off the land without properly transforming it for producing value. In doing so, they developed a narrative of surplus creation later revisited by the classical political economists favoring the rise of industrial capitalism. In tracing improvement discourses as central to the ecology of dispossession defining the dominant European discourses of being human in relation to the material world, I found it useful to turn to Karl Marx's account of the so-called primitive accumulation. In Part 8 of *Capital*, Vol. I., Marx's discussion of the processes that enabled the establishment of capitalism as a mode of production provides a powerful rejoinder to the dominant, idyllic narrative centering individual labor and rightful appropriation. The generation of surplus and the formation of world markets, he argued, was made possible by the systematic deployment of violence and dispossession. The European enclosures separated peasants from land, thus forcing them to rely on wage labor for survival and creating the proletarian workforce for industrial capitalism, a mode of production centered on exploitative exchanges between the owners of the means of production and the workers. Colonial resource extraction in the Americas and the enslavement of Indigenous and African populations fueled the rise of capitalism by satisfying the need of raw materials and labor, thus allowing the bourgeoisie to accumulate the initial surplus for revolutionizing production. For Marx "Conquest, enslavement, robbery, murder, briefly, force" and not individual industry and rightful appropriation, prepared the ground for the development of industrial capitalism (1976, 705).

As shown elsewhere in this book, a range of feminist, anti-racist, and decolonial scholars have addressed the limits of Marx's account of primitive accumulation while readapting his formulation for the analysis of the present. Often extending Rosa Luxemburg's perspective, this scholarship has rejected the notion of accumulation as a moment preceding the rise of capitalism and reconceptualized it as an ongoing process operating through the continuous dispossession and commodification of land, and the exploitation of gendered and racialized labor. For the purposes of this book, Marx's analysis remains relevant for countering improvement discourses of enclosures as rightful means of turning the earth into a source of value. Still more, the critical analysis of capitalist development described in vivid terms the drastic transformation of the earth occurring in the transition from feudal to capitalist social relations. This shift entailed what Marx and Engels in the *Communist Manifesto* (1848) related as the "Subjection of Nature's forces to man, machinery, application of chemistry to industry and agriculture, steam-navigation, railways, electric telegraphs, clearing of whole continents for cultivation, canalization of rivers, whole populations conjured out of the ground" (Marx and Engels 1976, 489). However, for Marx the technological reorganization of nature and livelihoods were part of the historical trajectory leading from capitalism to the liberation of living labor within the socialist society. To fully unpack this argument, and deepen the analysis of Marx discontinuities and overlaps with the hegemonic European modern model of the human, it is worth considering Marx's understanding of the role of labor with respect to nature and land.

Although Marx never fully developed the question of human existence in relation to nature, he returned to it often, thus inviting contemporary scholars to unearth rich ecological threads embedded in his *oeuvre* (Foster 2000; Saito 2017). Eco-socialist scholars, for instance, have highlighted Marx's insights into the capitalist plundering of the earth via commodity production. For this tradition, Marx understood society as existing within nature and the capitalist appropriation of the earth as producing a rupture, the "metabolic rift," in the interactions between human beings and the rest of the living.[7] Marx's materialism would provide a blueprint for envisioning modes of restoring earth cycles and the sustainability of human life within them through socialism as higher socioeconomic formation.

While the scholarship on the metabolic rift has provided generative accounts of Marx's salience for understanding socio-ecological questions, I join scholars who highlight Marx's more ambiguous understanding of the relationship between humans and the material world.[8] In a famous passage of *The*

Economic and Philosophic Manuscripts of 1844, Marx describes man as "species-being," a natural, conscious living being who manifests a peculiar mode of existence through sensuous activity. He writes, "The productive life is the life of the species. It is life-engendering life. The whole character of a species, its species-character, is contained in the character of its life activity; and free, conscious activity is man's species-character" (Marx 1988, 76). The concept of species-being returns in volume 1 of *Capital* (1867), where Marx offers a famous definition of labor as the process by which man "regulates and controls the metabolism between himself and nature" (Marx 1976, 283). He goes on to say that through this relation, man "develops the potentialities slumbering within nature, and subjects the play of its forces to his own sovereign power" (ibid., 283). Marx drew on scientific ideas of life as the constant transformation of matter. In using the notion of metabolism, he was inspired by the German chemist Justus von Liebig, a critic of industrialized agriculture who described exchanges and feedback within biological systems. Marx used metabolism to refer to the material exchanges activated by labor for the production and reproduction of human life. Now, it seems to me that the formulation of species-being, and even metabolism, reflects a process in which human beings act upon lifeworlds rather than in conjunction with them. Through labor, a form of energy capable of adding energy, man activates potentialities that would have otherwise remained latent. Human relationships to nature, therefore, can hardly be explained in terms of coevolution, as contemporary theorists of metabolism suggest. Rather, Marx describes the emergence of the human out of nature, as a living being capable of tirelessly mobilizing natural forces, animate and inanimate, for crafting its own transformation.

This distinction also emerges in a famous passage in which Marx argued that human labor's modification of the world is constantly helped along by natural forces: "Labour is therefore not the only source of material wealth, i.e. of the use-values it produces. As William Petty says, labour is the father of material wealth, the earth is the mother" (Marx 1976, 133–34). Why is Marx citing Petty's gendered metaphor here? He recognized Petty's work in the seventeenth century as foundational for a modern political economy, a framework that he critiqued by establishing an alternative account of capitalism's dynamics. But he seems to accept, even as he modifies, Petty's definition of labor's relation to nature. At first Marx posits a dual agency in the creation of wealth but then, by quoting Petty, he establishes an important difference between earth, that which has to be cultivated, and labor, that which transforms earth. While Marx brings to the fore the earth's role in creating wealth, he also undercuts it through the relation with human labor, associated with the masculine active principle.

What underpins Marx's understanding of the interactions between human labor and nature is the narrative of man as capable of transforming the world. As Donna Haraway puts it: "Of all philosophers, Marx understood relational sensuousness, and he thought deeply about the metabolism between human beings and the rest of the world enacted in living labor. As I read him, however, he was finally unable to escape from the humanist teleology of labor—the making of man himself" (Haraway 2008, 47). For Marx, the human species has a relation to nature by virtue of its capacity to use it as the ground for self-making.

The philosopher Jason Read (2003, 180) has suggested that the English translation of the German term *Gattungswesen* as "species-being" might be misleading in that it underscores biological meanings. He proposes that the French translation of *Gattungswesen* as la *vie générique* (generic life) might more accurately convey Marx's use of the term. This attempt to detach species-being from biology, however, overlooks how in Marx "generic life" indexes man's universality as opposed to animal particularity. Marx contrasts human species-being to the "species-life" of animals. Animal activity is identical to itself: It is purely instinctual and subordinated to physical needs. Humans, in contrast, can act and, simultaneously, confront the objects that they have created. Labor, or praxis, is the primary way through which human beings collectively transform nature and, by doing so, transform themselves. In the attempt to define what is proper to man as a laboring living being, Marx's species-being creates a distinction between the human and the nonhuman by which only the former acts upon the world, while the latter just exists.

Further, in the concept of species-being we find troubling echoes of the species discourse of the eighteenth and nineteenth centuries, one not only bound up with racialized and sexualized formations but also paradoxically connected to classic political economy's effort to naturalize capitalist relations of production. The scientific idea of the human as species was introduced in eighteenth-century Europe, where it was often conflated with race and used to naturalize the hierarchical ordering of biological differences. The development of species taxonomies was steeped in the colonial obsession for classification, connected to racial subjectification and infused with sexual difference. Carl Linnaeus's taxonomy is paradigmatic in this sense. The Swedish naturalist introduced the term *Mammalia* in the mid-eighteenth century to indicate the class of animals, including humans, characterized by the presence of mammary glands. Then he used the term *Homo sapiens* to distinguish between humans and other primates and defined four racialized subspecies ranging from the White, blond, and inventive *Homo sapiens europaeus* to the *Homo sapiens afer*, described as Black, lazy, and ruled by caprice. As the feminist historian Londa Schiebinger

(1993) has shown, the genealogy of *Homo sapiens* is not only highly racialized but also profoundly gendered. While Linnaeus used a female characteristic (the lactating breast) to emphasize the ties between humans and animals, he employed a traditionally male feature (reason) to indicate human uniqueness, or, more precisely, the uniqueness of the European White man.

Marx was not immune to the racialized legacy of species thinking. In *Grundrisse*, written between 1857 and 1858, he uses the distinction between species-life and species-being to contrast the Asiatic mode of production to the Germanic mode of production. Gayatri Spivak avers that Marx conflates the Asian individual with species life, natural life without human specificity. It is only with European feudalism and the movement toward urbanization in the Germanic mode of production that the self-reflexive relationship with nature typical of species-being emerges. Spivak notes that in Marx's description of the Asiatic individual "it is almost as if Species-Life has not yet differentiated itself into Species-Being" (Spivak 1999, 80). The species distinction is now recast in historical as well as geographical terms.

In *The Order of Things*, Michel Foucault (1970) argues that modern man emerged at the intersection of three discursive domains—life, labor, and language—articulated by biology, political economy, and linguistics, respectively. These are interdependent domains, characterized by an intense flow of ideas. Political economy, for example, borrowed heavily from the species taxonomy developed by natural history. Adam Smith, who was familiar with the work of Linnaeus, proposed the market as a natural, self-regulating force independent from individual agency and able to guarantee the perpetuation of the species against extinction (Schabas 2005; Cohen 2013). Political economy had an anthropological foundation insofar as it constituted itself in relation to "the biological properties of the human species" (Foucault 1970, 257). Marx's project countered political economy's attempts to naturalize an economic order grounded on private property and the exploitation of labor. Yet by thinking about labor as species capacity, he imported from classic political economy the idea that labor is what makes us human.

Marx offers crucial insights for analyzing the modern liberal ecology of dispossession. Documenting the violent dissolution of the commons, and pointing toward the global geographies of accumulation, he challenged dominant accounts of rightful property formation and improvement. Yet, he understood labor as the key source for the transformation of nature in the historical process leading to the development of capitalism and beyond. In this respect, his work presents important continuities with the modern tradition that invests the human with creative powers over the material world. By privileging an anthropology of the production of man by man, he also laid the ground for contemporary Marxist approaches to the commons that I discuss in Chapter 3.

Conclusion

The historical trajectories of the commons presents considerable ambiguities when retold in the context of what I define as the transatlantic ecology of dispossession, a contingent socio-ecological formation produced at the intersection between the consolidation of market and monetary economies, the rising conception of the material world as a site of scientific exploration and value extraction, and the expansion of the colonial project. This arrangement of dispossession, with its attendant definition of what it meant to be human, emerged through uneven encounters with alternative modes of inhabiting the earth and played out differently in the enclosure of the European commons and the appropriation of Indigenous land in the colonies. Predicated upon the liberal assumptions about humanness, labor, property, and land articulated by the English improvers, the ecology of dispossession acquired a distinctive racialized and gendered dimension by stratifying human beings in a matrix of difference. White, masculine property owners and settlers embodied the hegemonic model of the human entitled to appropriate other peoples' land and bodies (Harris 1993; Bhandar 2018). The commoners living in the English countryside were classified as lazy and unable to improve land and make it profitable. At the same time, via transatlantic routes, Indigenous modes of inhabiting the earth were racialized as belonging to a vanishing state of nature to be superseded by rightful colonial appropriation, and African people were racialized through a process of "thingification" (Césaire 1972) that disentangled them from their land through kidnapping and enslavement. Improvement, in other words, required what Stefano Harney and Fred Moten call speciation, the operation through which the self-owning European man "extracts and excepts himself from the earth in order to confirm his supposed dominion over it," while also forcefully setting himself apart "from other groups—particularly, fundamentally, in violent speciation, from groups that do not own (either self or earth)" (Harney and Moten 2021, 29). The process of forming and distinguishing the human, the less-than-human, and the nonhuman created differential rather than equivalent modes of dispossession. For the improvers, the commons and the peasant economies of subsistence were a relic of the past standing in the way of historical progress. Yet, in the Americas aspects of shared land use were mobilized by settlers in the process of claiming ownership of Indigenous land. In tracing the tense relationship between the commons and the modern ecology of dispossession this book pays attention to both radical conflicts and uncomfortable convergences. Even Marx, a prominent critic of the liberal ecology of dispossession, seems to share some of its underlying assumptions, namely the idea of the commons as belonging to the precapitalist past, and the narrative of the human as self-making being, transforming the material world through the power of labor.

PART II

PRODUCTION, REPRODUCTION, CARE

3

ENGAGING POTENTIALS AND LIMITS OF THE MARXIST COMMON

> "If you don't admire something, if you don't love it,
> you have no reason to write a word about it"
> —Gilles Deleuze (2004)

Over the past two decades the Anthropocene has become one of the most debated terms in the geosciences and environmental studies, rapidly moving beyond academia into art galleries, journalism. and popular culture.[1] This concept portrays the human species as the dominant geological agent provoking a dramatic perturbation of earth's processes and generating environmental changes including global warming, ocean acidification, and biodiversity loss. Within the critical environmental humanities and allied fields, a robust body of scholarship in feminist theory, critical race theory, and Indigenous studies has argued that the universal *anthropos* featured in the prevalent scientific narrative of the Anthropocene obscures uneven responsibilities and vulnerabilities in creating the conditions for the current planetary problems. The generality of the anthropos, in other words, masks the particularity of a specific model of the human, the White man of European modernity entitled to appropriate the material world in the quest for capitalist improvement and, in the process, turn the earth into resource and sink for waste (Todd 2015; Haraway 2016; Tola 2016; Di Chiro 2017; Vergès 2017; Pulido 2018). The Anthropocene narrative perpetuates what Sylvia Wynter (2003, 263) has described as "the overrepresentation of Man," that is, the universalization of the Western dominant conception of the human and the dehumanization of other ways of being. This term, in other words, leaves in the background the histories of colonial, racial, and economic violence that for a long time have shaped the experiences of people and places outside the Western world (Danowski and de Castro 2016; Povinelli 2016; Ferdinand 2021). At the same time, much Anthropocene discourse carries a salvific dimension, foregrounding technoscience as the means for overcoming existential challenges to human life on earth. In the face of unraveling ecological crises, the anthropos is presented has a geological agent remaking itself and reconstructing the earth (Neyrat 2019). While these critical perspectives on the Anthropocene do not entail a wholesale dismissal of this concept, they raise the question of what politics might be pursued within and against prevalent narratives that foreground an undifferentiated human species capable of simultaneously causing and remediating compounded environmental challenges. It is against the backdrop of these debates that this chapter continues to examine the resurgence of the commons in social theory and activist movements.

Theoretical in scope, this chapter explores the conceptual promises and limits of contemporary Marxist contributions for reimagining the commons beyond the Anthropocene.

My focus here are debates primarily associated with the Italian tradition of autonomist Marxism that analyze recent shifts in the global political economy. Specifically, I consider how, since the 2000s, some prominent scholars in this diverse tradition have been defining "the common," in the singular form, and its role in the rise of "cognitive capitalism." From this perspective, the common indexes, broadly speaking, the cooperative capacities of workers within a system of accumulation in which knowledge, information and communication are the key sources of value.[2] Unlike Elinor Ostrom (1990), who understands the commons, in the plural, as a stock of resources that are collectively managed and as a form of governance that coexists alongside state and market arrangements, autonomist Marxists are interested in the common as a mode of production that has the potential to exist outside of capitalist relations (Giuliani and Vercellone 2019).

In revisiting influential formulations of the common, namely Paolo Virno's philosophical meditation on "human nature," and Michael Hardt and Antonio Negri's emphasis on the biopolitical production of the common, I am interested in assessing their accounts in the context of the unfolding ecological crisis and the intertwined intensification of racial, gendered, and environmental violence. In many ways their writings were my entry point for reimagining the commons and I deeply value their investment in thinking through and with political struggles and activist movements in Europe and elsewhere. As I have been grappling with it throughout the years, I have found their conceptualizations of the common generative and yet frustrating. This ambivalent relationship has been shaped by my upbringing in Italian political networks deeply intertwined with the autonomist tradition through various generations of scholarship and activism. However, encounters with feminist perspectives and political ecologies have helped me to see the limits of autonomist thought in a time of socio-ecological exhaustion. To articulate this ambivalence, I engage in the close reading of Virno, Hardt, and Negri, showing how their writings disrupt capitalist formations of the human and yet propose models of *Homo sapiens* and *Homo faber* that intersect the Anthropocene discourse of human exceptionalism. This chapter asks: How does autonomist Marxist formulation of the common rely on or challenge the figure of the anthropos? What does this approach offer, or not offer, for making common(s) in ways that make explicit human dependence on the more-than-human world?

There is little doubt that autonomist Marxism has had a profound impact on the resurgence of the common(s) as a project generating alternatives to capitalist social relations. Central to this line of thought is the focus on commoning as a collective human activity, as relational and affective labor that is at the same time an indispensable engine for capitalist innovation and the process through which creating alternatives to capitalism. However, Italian autonomist Marxists have largely remained silent about how their ideas relate to the ecological devastation and its unevenly distributed effects. In a spirit of convivial critique, I highlight their insights about the common in current regimes of value extraction and anti-capitalist organization even as I register their attachment to the productive human that is at the core of the Anthropocene.

This chapter opens by briefly contextualizing Negri's and Virno's trajectory from Italian workers' movements in the 1960s and 1970s and introduces the notion of labor's autonomy in relation to environmental problems. Then I move on to discussing the centrality of human cognitive and relational faculties in Paolo Virno's analysis of the capitalist restructuring. I examine his use of Gilbert Simondon's theory of individuation and suggest an alternative reading of Simondon that opens up the space for imagining a more-than-human common. Next, I extend Michael Hardt and Antonio Negri's gesture toward an "ecology of the common" moving beyond their distinction between natural and social commons. Throughout, this chapter explores how to build on autonomist Marxism's investment in the power of ever-shifting collectives without embracing the human as the central agent of commoning. The analysis of this body of work allows me to interrogate influential conceptualizations of the common and their political potential at a time of ecological crisis. There are practical and theoretical reasons for such a project. The multiplication of social and environmental breakdowns reflecting the imbricated histories of capitalism and colonialism requires profound shifts from modes of inhabiting the earth as theater of human transformation. What might be needed is not a project of commoning driven by a reinvigorated human but

Engaging Potentials and Limits of the Marxist Common

one that politicizes the attachments between human collectives and the socio-ecological conditions of existence.

CENTERING LIVING LABOR

Both Negri and Virno were members of the Italian revolutionary organization Potere Operaio (Workers' Power) until 1973, when the organization dissolved into the broader movement of Autonomia. They were active in the cycle of struggles that began in the 1960s and culminated in 1977 with the irruption of new subjectivities in the Italian political scene that simultaneously expressed the refusal of work discipline in the industrial factory and the invention of new modes of living. As defendants in the "April 7th trial," Negri and Virno spent time in prison, accused of subversive association and armed insurrection. Incarcerated for four years with charges of masterminding terrorist activities, Negri ended up escaping to Paris where he was able to join the French academia.[3] These activist scholars have been leading voices in Marxist debates on the massive shift in the nature of labor and political organization occurring since the 1970s and often described as the transition from Fordism to post-Fordism. These terms mark the discontinuities between the rigidities of Fordist mass-scale production systems and labor markets, and the post-Fordist regimes of flexible accumulation marked by the outsourcing of production and the technological reorganization of supply networks.[4] What distinguishes the autonomist approach, also known as *operaismo* (workerism), is the emphasis on the role of workers' struggles as preceding the process of capitalist restructuring. From this perspective, the transition to post-Fordism was the capitalist response to workers uprisings occurring from 1968 to the late 1970s.

From the outset, autonomist thought has been shaped by the close relation between intellectuals and factory workers built through continuous exchanges and shared struggles arising from the collaborative analysis of factory floor dynamics. Porto Marghera, a large petrochemical complex lying across the water from Venice was a strategic pole of the process that turned Italy from a rural economy into a world industrial powerhouse. Here, in the 1970s, Italian workerism experimented with a distinctive style of political intervention and knowledge production that blurred the boundaries between political theory and practice. This political laboratory produced concepts and experiences that left lasting marks on autonomist theorizing. What is interesting is that workers' collectives in Porto Marghera also drew attention to socio-ecological questions. They developed a broad critique of capitalist discipline that also engaged the toxicity of petrochemical manufacturing for human bodies and environments. Through wildcat striking, occupations, marches, and political proposals they turned public health and environmental degradation into terrains of political contestation, crafting alliances with health care professionals, scientists, and local communities. As relevant scholarship has demonstrated, this was a striking example of *worker environmentalism*, developing an antagonistic critique of capitalist technology *and* the human and environmental costs of industrial modernization (Barca 2012; Leonardi 2019; Feltrin and Sacchetto 2021). The ecological aspect of workers' struggles, however, has left relatively few traces in contemporary accounts of the common.

Negri has often fondly recalled the mobilizations in Porto Marghera as foundational for the emergence of *operaismo*. For instance, in a video interview with Ed Emery from 2022, he describes them as an engine in the modernization of Italy leading the country out of the backward condition of peasant life, on a path of development marked by workers' struggles for the liberation of productive forces.[5] This account of the story tends to overlook the environmental dimension of this struggle to focus on the conflict between the working class and capitalism as a significant instance of modernization. It foregrounds human living labor as a progressive force rejecting its own commodity status while disjoining it from the material possibilities of life, from the land, the water, the skies, and the urban environments that industrial capitalism turned into sites of toxic manufacturing and sinks for waste. However, as Sara Nelson and Bruce Braun note, the rapid transformation of the Italian economy in the 1960s was a product of the simultaneous mobilization of labor power employed in the factories and the "concomitant intensification of *nonhuman* productivity" (Nelson and Braun 2017, 228) starting from fossil fuel's dead ecologies fueling industrialization. In other words, it required not just the exploitation of workers but vast operations for extracting the "buried legacies of plant-life" (Huber 2013, 12). In recalling how Negri overlooks the ecological dimension of workers' struggles in Porto Marghera, I highlight a habit of thought that presumes the autonomy of labor not just from capitalism but also from its material conditions of possibility. This is relevant for the purposes of this book as the assumptions about human living labor within Italian workerism are at the core of influential approaches to the common.[6]

To further elaborate on the concept of autonomy and its significance for thinking about the commons, it is worth briefly examining how it was developed through an original interpretation of Marx's "Fragment on Machines." Part of the *Grundrisse*, the "Fragment" is the key text autonomist Marxists draw on to make sense of the shifting relationship between labor and capitalism. Here Marx reflects on the relationship between living labor and the dead labor objectified in machinery and technology. He suggests that capitalism increasingly depends on the "general intellect" (Marx 1973, 706), that is, the accumulation of social knowledge objectified by capital in technical machines. Autonomist Marxists have used the concept of general intellect to analyze the shift from Fordism, the standardized model of factory production typical of advanced industrial capitalism, to post-Fordism, a flexible regime of accumulation connecting heterogeneous sites of production and forms of precarious labor. Their distinctive reading of the general intellect privileges living labor as that which is only ever partially captured by capitalism. According to this analysis, largely rooted in the post-1977 Italian landscape of repressed insurrection and capitalist restructuring, capitalism has converted the refusal of factory discipline expressed by new antagonistic subjects into productive activities that blur the boundaries between labor and life. By and large, post-Fordist workers are no longer required to perform repetitive tasks. What is now put to work are human beings' communicative, affective, and relational capacities. As Paolo Virno (2004) puts it, "in Post-Fordism, the general intellect does not coincide with fixed capital, but manifests itself principally as a linguistic reiteration of living labor" (ibid., 106). In this regime of accumulation, communicative and cognitive skills are at the very heart of production.

If Marx identified the general intellect with the abstract knowledge subsumed by the machines, autonomist Marxists argue that "general social knowledge" cannot ever be fully integrated within fixed capital because it is "actually inseparable from the interaction of a plurality of living subjects" (Hardt and Virno 1996, 194). The general intellect results from the interrelation of living labor and the fixed capital of technologies. But human communicative and relational capacities constitute the driving force of the productive process. Thus, although incorporated into technology, the general intellect is also continuously renewed through the interaction of a heterogeneous multitude of workers. In this sense, it remains at least partially autonomous. The idea that the post-Fordist regime of production extracts value from the social cooperation that underpins the production of ideas, codes, and images is variously articulated in autonomist circles. In the next sections, I examine its implications for contemporary imaginings of the common through a close reading of the work of Paolo Virno, Michael Hardt, and Antonio Negri as well as other autonomist thinkers.

THE RACE AND GENDER OF SPECIES-BEING

Virno's philosophical approach to the common stands out for its persistent investment in human nature, a notion somehow out of sync with feminist and anti-racist contestations of who has come to count as human in Western modernity. Although Virno has not directly engaged the Anthropocene, the anthropos is at the core of his analysis of post-Fordism. This form of accumulation, he argues, mobilizes the biolinguistic faculties that set *Homo sapiens* apart from the rest of the living. These faculties, understood as inexhaustible potentiality rather than as a timeless given, constitute the common of humanity, what might be actualized in the form of "engaged withdrawal" from capitalism and the state (Hardt and Virno 1996, 196). In works such as *A Grammar of the Multitude* (2004), *When the Word Becomes Flesh* (2015), and *E cosi via, all'infinito* (2010), Virno attempts to reconnect the history of labor with natural history, the transformation of social relations with the powers of the human as natural being. At the intersection between the human form of life and the post-Fordist transformation, he contends, new modes of being together as a common may emerge. As the key thinker of the "naturalist" tendency within Italian autonomism, Virno offers a compelling point of entry for exploring the limits and possibilities of autonomist Marxism for thinking about the common in the Anthropocene. In exploring Virno's approach to the common, I ask: What is the anthropos for Virno? How does it intersect with the hegemonic model of the (White) man?

Virno's wager is that contemporary capitalism produces value by harnessing the "biological invariant" common to human individuals: the potentiality of speech and relationality. While other animals dwell in a fixed environment that triggers specialized behaviors, *Homo sapiens* is characterized by innate disorientation (*disambientamento*). The lack of specialization, "the habit of not having solid habits," translates into a fundamental oscillation between blockage and innovation, negation and affirmation (Virno 2005, 29). Here Virno draws on philosophical anthropology's attempt to compare man and animal as a way to grasp the distinctive traits of man. Influential in Germany between the 1920s and 1950s, the philosophical anthropology of Helmut Plessner and Arnold Gehlen was indebted to Jakob von Uexküll's (2010) ethological study of the relations between organisms and their *Umwelten*, lifeworlds defined by correspondences between sensory capacities and environmental forces. Uexküll, however, seemed inclined to think that humans, too, act within a particular milieu, one more complex than that of many other living beings and yet functioning on the basis of similar operating principles. In contrast, philosophical anthropologists argued that the human species is fundamentally deprived of *Umwelt* and therefore compensates for this deficiency through the creation of cultural environments and the capacity for self-reflexivity. Virno and philosophical anthropologists agree that all organisms are enmeshed in lifeworlds. But humans, they contend, are eccentric beings, deprived of a milieu and therefore at a distance from themselves. The lack of a fixed environment and radical openness to the world sets *Homo sapiens* apart from other organisms. As beings that do not fully coincide with their milieu, humans have the capacity to transform their form of life. Now, it is precisely because post-Fordism relies on human nonspecialization, that it engenders, according to Virno (2009), a historical and social repetition of anthropogenesis. In other words, the post-Fordist organization of labor corresponds to an ontological condition that oscillates between repetition and the capacity to invent the new.

It is important to note that when Virno draws attention to the "since always" of human nature he is evoking not a transhistorical essence but a potentiality that is immanent in human beings. He is interested in how the "right now" of post-Fordism, with its insistence on flexibility and precarity, forces a reconsideration of the human as species. In this respect, his intervention partially overlaps with Dipesh Chakrabarty's (2009) point that the Anthropocene "requires us to put global histories of capital in conversation with the species history of humans" (ibid., 212). For both thinkers, it is not that the human has a species destiny to fulfill but that the current global situation imposes a return to species thinking. What is perplexing, however, is the conflation between human generality and global dynamics. Chakrabarty links the global fact of anthropogenic climate change to the return to the generality of the species. In Virno's analysis of the transformation of global capitalism, natural history is conflated with the history of *Homo sapiens*. In both cases what remain unexplored are historical stratifications along the axes of race, gender, sexuality, and geography that render human beings differentially exposed to the precarization of life. Further, these scholars do not take into account the other-than-human forces that enable, and disable, human existence and that capitalism variously appropriates and enrolls in productive processes (more on this in Chapter 4). According to Virno, in the context of post-Fordist transformations, Marx's category of *Gattungswesen* (species-being), the generic existence of humanity, acquires new relevance. He writes, "Roles and tasks, in the post-Ford era, correspond by and large to the *Gattungswesen* or 'generic existence,' which Marx discussed in *The Economic and Philosophic Manuscripts* of 1844" (Virno 2008, 78). We have come full circle: Human nature is the point of integration between historical materialism, the critical trajectory that began with Marx and connects productive forces and social relations, and "naturalistic materialism," by which Virno means the investigation of the distinctive capacities of the human species.

As I showed in Chapter 2, Marx's species-being foregrounds man's capacity to act upon and activate nature's potentialities. But while Marx engages the relationship between society and the material world and the ways in which value is created through exchanges with nature, Virno conceives the constitution of the human, what he calls anthropogenesis, as strangely isolated from the larger milieu of natural forces. Through his emphasis on human nature, Virno reminds us that as a species lacking a particular ecological niche, we humans are part of nature and that our nature is inseparable from specific historical articulations and relations of production. Yet, the rest of the living and nonliving, animals, plants, and inorganic matter appear as the mere backdrop of human natural history as if this was the history of just one species and not also of other forms of life and nonlife within the earth system. Jason Read observes that for Virno "human nature is its history" (Read 2017, 265). But Virno's analysis of the capitalist exploitation of human nonspecialization as a species does not engage the question of capitalist reproduction through the exploitation of labor and the appropriation and dispossession of people and land. His account of post-Fordism as a historical reiteration of anthropogenesis risks producing a double foreclosure. On the one hand, it elides the histories of appropriation of what Jason Moore (2015) calls "cheap natures." These include not just the appropriation of diverse forms of unpaid and forced labor, but also the cheap energy extracted from inorganic natures and the cheap food made available through agricultural revolutions leading to the exhaustion of labor and land. On the other hand, Virno's account of anthropogenesis obscures the effects of racialization and feminization that the species discourse has historically both enabled and entailed. As I recalled in the final section of Chapter 2, eighteenth- and nineteenth-century species discourses were deeply entangled with colonial formations of race. As Achilles Mbembe (2001) points out, the colonial "grammar of animality" excluded Black people and natives from the field of the human in ways that then justified "the domestication of the colonized individual" (ibid., 236). This grammar figured animals as beings that could not distinguish between themselves and the external world. Similarly, colonized people were seen as unable of transcending biological life and therefore animalized.[7] The emergence of *Homo Sapiens*, the species of which Virno considers the evolution within advanced capitalism, was imbricated with these racial stratifications.

In Virno's work, the foregrounding of labor as potentiality immanent in the whole of humanity ends up bypassing the potentialities of the ecological and geological milieu that provides the conditions for what "we" have come to understand as human. Moreover, it unwittingly reproduces narratives of human autonomy without questioning their colonial and racialized legacies. Operating within an updated workerist framework, Virno falls short of providing a counterpoint to the narratives of the Anthropocene that posit "generic" man as the primary locus of geopolitical agency. Even as I emphasize these limits, it is also worth noting that Virno offers an expansive and nuanced notion of the collective that displaces the political ontology of modernity, particularly the idea that the political community is made up of individuals who have left behind the state of nature. It is to this question that I turn in the next section.

STATES OF NATURE

Virno's insistence on the political valence of human nature in the present context of capitalist accumulation poses an important challenge to the tradition of European modern liberal thought that conceives of isolated individuals as the basic unit of social life and politics. For example, in the work of Thomas Hobbes, one of Virno's favorite targets, the relationship between the many and the sovereign is unidirectional. It begins with a multitude of hostile individuals scattered in the state of nature and culminates with their submission to the law in exchange for protection from violence and death. Through the transition from the prepolitical state of nature to the civil state, the multitude becomes the people, an aggregate of individuals whose interests are mediated by the universal figure of the state. Hobbes presents the state of nature as a state of individuals at war with each other. John Locke, as I discussed in Chapter 2, proposes a narrative of the individual that asserts its full humanity through labor and appropriation. Both Hobbes and Locke associate the establishment of civil society to the overcoming of the state of nature. Building on Marx, Virno (2010) provides an antidote to the Western modern ontology of the human as competitive and possessive individual. For the Italian philosopher the state of nature coincides with the common, that is, the shared linguistic faculty of the human species that renders human beings constitutively open to social relations. While the universal of modern thought results from the abstraction of recurrent elements that return in a number of already individuated entities, the common provides the conditions for the emergence of a multitude of singularities. There is no dividing line between the common and the multitude, only temporary trajectories of dislocation that continuously renew the common. This means that there is no overcoming of the state of nature, only countless realizations of its potentiality. What I find interesting is that Virno posits the common as natural formation and as the ground of political life. This is a fascinating proposition and one that marks a rupture with the dominant Western modern conception of the human. However, it presents the limit of understanding nature, and the common, exclusively in term of human biolinguistic faculties, that is, as the nature of the anthropos. This is even more perplexing if one considers that Virno's philosophy of individuation has been influenced by Gilbert Simondon, a thinker who has challenged the primacy of the anthropos in Western critical thought. While an in-depth analysis of Simondon is out of the scope of this chapter,[8] I want to take a closer look at him to show that his work allows the reworking of the common beyond the anthropos, as a form of collective individuation capable of cultivating the attachments to the living and nonliving forces that constitute its condition of possibility.

Simondon has been largely interpreted as a philosopher of technics and technogenesis (MacKenzie 2002; Stiegler 1998). Explicit references to politics in his work are sparse to say the least.[9] Yet, the relevance of his model of ontogenesis for elaborating alternatives to the modern fixation with individuals as the basic unity of social and political life has become the subject of a lively debate.[10] Instead of focusing on elementary units or essences, Simondon shifts attention onto *ontogenesis*, that is, the process through which specific forms of life come into being and change over time. Ontogenesis originates in a metastable "preindividual reality," which Simondon, inspired by pre-Socratic philosophers, also calls *nature*. In physics and chemistry, metastability indicates a system in a state of tension that even the smallest disturbance can alter. The preindividual is characterized by a level of potential energy and internal incompatibilities that trigger a change in the system, leading to the emergence of more or less completed individuals. Individuation takes place when a communication is established between different orders of magnitude that coexist within the metastable system. This produces a new phase of being, a medium order that provisionally resolves an internal problematic. The growth of a plant is an example of ontogenesis: "a vegetable institutes a mediation between a cosmic order and an infra-molecular order, sorting and distributing the chemical species contained in the ground and in the atmosphere by means of the luminous energy received from the photosynthesis" (Simondon 2009, 16).

Simondon describes the dynamic of differentiation within the preindividual as transduction, an operation that cuts across the physical, the biological, the social, and the technological to express individuals that remain in relation to the metastable field. The same operation of transduction produces living and nonliving individuals, thus destabilizing the oppositional hierarchy between life and nonlife, organic and inorganic, human and nonhuman. It affects not just individuals but also the preindividual milieu by transforming it in a way that does not impoverish its potential to engender endless variation. Further, Simondon connects psychic and collective individuation through the concept of transindividual. He suggests that transductive operation puts into communication charges of preindividual reality that are shared by individuals. The transindividual, in other words, is the surplus of potential that is actualized when a subject enters collective individuation, when a being is affectively connected to what in oneself is more than individual. Simondon understands individuation as a more-than-human relational process of mutation. Individuation is not human to begin with; it emerges out of an inhuman milieu and unfolds in innumerable directions. This does not mean denying human singularities but invites moving beyond bounded notions of the human as autonomous from animal, plant, mineral, and technological existence. It seems to me that this has important implications for thinking of the common as a process in which the human exists through relations of becoming with the material world that exceed the logic of resource appropriation and extraction.

Virno, however, glosses over Simondon's insistence on the preindividual as a prevital field of disparation that propels innumerable trajectories of life and nonlife. Instead, he uses the preindividual to describe the common potentialities of the human that are put to work in the circuits of post-Fordist accumulation. Contemporary forms of labor mobilize "the most universal requisites of the species: perception, language memory and feelings" (Virno 2004, 77). At the same time, these biolinguistic human capacities are continuously renewed through collective individuation: "only within the collective, certainly not within the isolated subject, can perception, language, and productive forces take on the shape of an individuated experience" (Virno 2004, 78–79). The multitude, an unstable network of cognitive workers, is the name for a form of collective individuation in which the many persevere as many and always carry within themselves shares of preindividuality. It is in the multitude that a second dimension of the common may emerge: "Besides being preindividual, it is transindividual; it is not only the undifferentiated backdrop, but also the public sphere of the multitude" (Virno 2009, 64). This implies that the common for Virno indicates more than the human linguistic and relational faculties that capitalist accumulation organizes in processes of value extraction. It also indexes the possibility of politicizing linguistic and relational practices in a process of engaged withdrawal from capitalism, that is, the creation of alternative forms of existence. Virno is careful not to characterize the multitude simply as a network of rebellious singularities capable of creating postcapitalist modes of living. It is a much more ambiguous formation, one that reflects the ambivalence of *Homo sapiens*. Importantly, this assessment of the multitude underscores the indeterminacy of any radical political project.

Even as I appreciate Virno's nuanced conceptualization of the multitude, I remain unconvinced by his political reading of Simondon in which the process of individuation that might actualize the common begins and ends with the anthropos. In contrast, I understand the preindividual as that which displaces the centrality of the anthropos and look at the common as a project that requires the interplay of disparate beings and forces, not all of which are human. Simondon's preindividual does not coincide with human nature but is closer to what pre-Socratic philosophers called *physis*. The philosophy of ontogenesis revitalizes *physis* and rejects its division from *techne*, what emerges out of nature and what is produced by human activity. The preindividual makes individuation possible but it is not reducible to any particular trajectory of becoming, including the human. Unlike much of modern Western thought that understands the social as processual and dynamic, capable of mobilizing a malleable nature, the ontogenetic approach frames preindividual nature as what creates the conditions for the production of variations that reverberate through the social. As Muriel Combes (2013) suggests, Simondon poses a striking question: "What can a human do insofar as she is not alone?" (ibid., 50). I am interested in reimagining the common through this question, not as a universalizing feature of the species but as a situated political project contesting modes of value extraction that target bodies and their conditions of existence, what allows them to breath, feel, and act together.

Virno does away with the notion of politics as an overcoming of the state of nature deeply ingrained in the liberal tradition. In thinking the common, he connects natural potentialities with a politics that is entangled with the development of the forces of production. This is a powerful move, but one that presents the limit of analyzing the human species as a rather undifferentiated aggregate of living beings and in utter isolation from ecological and geological formations. Simondon, on his part, does not provide an analysis of power, an understanding of how particular individuations of preindividual tensions come to acquire quasi-stability as abstract models with violent effects on particular categories of bodies. He provides little insight on how gender, race, class, and species have become hierarchical categories producing distinctions within the human and between human and nonhuman beings. What Simondon offers, however, is the forsaking of anthropology as the ground of politics. This, I contend, does not mean to do away with politics altogether. On the contrary, it poses the challenge of cultivating different forms of politics. Instead of thinking of preindividual nature as the mute substratum that is left behind in the human process of collective becoming, Simondon calls attention to the indeterminacy of physis that makes politics possible. What is at stake here is the opening up of an approach to politics, and the common, that does not lose sight of the prevital and living elements that are elaborated through psychic and collective individuation. As that which creates the conditions for trajectories of becoming, preindividual nature "renders social transformation thinkable" (Combes 2013, 54).

As Alberto Toscano (2007b) aptly notes, Virno's thinking of the preindividual common as human nature implies that a new social configuration lies in a state of latency, as if waiting for the propitious convergence of anthropogenesis and capitalist development to emerge. From this perspective politics would consist in the engaged withdrawal of human linguistic capacities from capitalist control. Simondon instead gestures toward a politics that begins with "the invention of a communication between initially incompossible series; as invention of a common that is not given in advance" (ibid., 3). Moving along these lines, I argue for a reconsideration of the preindividual as a more-than-human field of potentialities, the ground for the difficult task of making commons. Neither the reservoir of human linguistic faculties nor its actualization beyond capitalism, the commons could be thought of as a project enacted by humans as beings "with and of the earth" (Haraway 2016, 60).

Virno offers a profound rethinking of the relationship between natural human potentials, their historical realizations, and political relevance for the present time. By doing so, he unsettles one of the key tenets of Western political thought, namely, the idea that politics begins where the realm of nature ends. Yet, as this chapter demonstrates, he also conflates natural history with the history of the laboring human. For Virno, as for Marx and much of autonomist thought, man produces man, a figure whose only attachment is to himself. This chapter focuses on different genealogies of the human and the common. It attends to the racialized and gendered logic that has historically informed *Homo sapiens* as the dominant model of man that acts upon and transforms the world. In doing so it problematizes the investment in the undifferentiated anthropos as a prime mover of history. At the same time, building on Simondon, it explores instances of the commons capable of making present the other-than-human forces operating "within everything we think is ours, or our own doing" (Sharp 2011, 9). The capitalist Anthropocene is replete with assumptions about *Homo sapiens* as agent of catastrophe and source of salvation. Radical thought ought to operate within and against these assumptions to illuminate at once the larger milieu of beings and forces that make the human possible while also paying attention to the uneven histories of violence that stratify living labor.

BIOPOLITICAL COMMON

Michael Hardt and Antonio Negri's influential work on the commons diverges from Virno's interest in human nature and focus on the biolinguistic abilities of the species. At times, they seem more aware of pressing ecological questions and yet, like Virno, they identify the common with living labor and insist on its potential autonomy from capital. In *Commonwealth* (2009), the volume that concluded the trilogy begun with the best-seller *Empire* (2000), Hardt and Negri present a compelling formulation of the common. They suggest that an "ecology of the common" would focus "on nature and society, on humans and the nonhuman world in a dynamic of interdependence, care and mutual transformation" (ibid., 171). Ten years later, in an article taking stock of the aggressive resurgence of national sovereignty, the digitally-enabled forms of value extraction, and the restructuring of global governance, they make a very similar remark. They refer to Indigenous struggles that pose "the need for humans to establish a new relationship with the earth, characterized by relations of interdependence and care—to make the earth common" (Hardt and Negri 2019, 83). These are intriguing propositions but, as I will show, they remain largely unattended in Hardt and Negri's writings. It is as if the ecologies of the common are invoked but not fully developed in an expansive framework that foregrounds the insurgent potentialities of social cooperation in the global order of capitalist accumulation. Even as Hardt and Negri address the earth common, they privilege the analysis of capitalist production and global governance without attending to the close entanglements between these and the material dynamics of extraction, depletion, pollution, and waste. In other words, the social and the ecological dimensions of the common remain largely distinct and the former is prioritized over the latter. Complicating this approach, this book seeks to contribute insights for addressing the socio-ecological conditions of possibility of the common.

In this section I consider Hardt and Negri's concepts of biopolitics and living labor, and I explore their relationship to the ecology of the common. I draw primarily on the *Empire* trilogy, but also on Negri's earlier writings and Hardt's solo work. I show how Hardt's and Negri's politics of the common and the multitude oscillates between Gilles Deleuze and Félix Guattari's transversal conception of life and the Marxist allegiance to human self-making activity. I suggest that ultimately, their social ontology of living labor as the human capacity of transforming the world dissolves nature into the social. Hardt and Negri expand on Michel Foucault's meditation on biopower and biopolitics[11] to analyze the qualitative transformation in the relationship between capitalism, labor, and the production of subjectivity. Drawing on Baruch Spinoza as well as Deleuze and Guattari, they contend that biopolitics consists of the human creative capacity to resist capitalist biopower through the production of new subjectivities and forms of life.

This emphasis on the power *of* life is connected to a set of other concepts that Antonio Negri has consistently developed since the 1970s: living labor, *potentia*, constituent power, multitude. These are all ontological figures in a framework in which ontology is understood not as a myth of origin but as a process of metamorphosis and becoming of the social, as the human capacity of constituting and continually reinventing itself. Already in *Insurgencies*, the book first published in 1992 that anticipates many themes of the *Empire* trilogy,

Negri inscribes Foucault's biopolitics in this theoretical constellation, claiming that:

> after demonstrating how power can subjugate humanity to the point of making it function as a cog of a totalitarian machine, . . . Foucault shows instead how the constitutive process running through life, biopolitics and biopower, has an absolute (and not totalitarian) movement. This movement is absolute because it is absolutely free from determinations not internal to the action of liberation, to the vital assemblage (agencement) (Negri 1999, 27).

This ontological notion, which commentators such as Alberto Toscano see as a departure from Foucault's anti-universalist biopolitics,[12] is clearly influenced by Deleuze's interpretation of Foucault. Deleuze suggests that Foucault's thought culminated in a "certain vitalism" (1988, 93) that understands life as the capacity for resisting force. Deleuze's take on biopower develops against the backdrop of his concept of life as an impersonal force that cuts across the boundaries of subjects, organisms, and species. For Deleuze and Guattari, life runs between the immanent fields of the natural and the social to create provisional convergences, assemblages of heterogeneous elements that constitute processes of production. The breaking down of the barriers between nature, the social, and the subjective is crucial in this framework. Nature is not the substratum of a human enterprise but a process of production in itself. It is not ontologically separated from the social; they are part of the same plane of immanence: "they are one and the same essential reality, the producer-product" (Deleuze and Guattari 1987, 5).

Instead of fully addressing Deleuze's challenge to think through the natural and the social together, Negri turns to the workerist reversal of the relationship between capital and labor, and makes Marx's concept of living labor the central category for understanding the power of life. In contrast with the longstanding argument that capital is the driving force of the process of production, workerism claims that living labor and workers' struggles compel capital to transform. Hardt and Negri develop the same argument when they contend that:

> The history of capitalist forms is always necessarily a *reactive* history. . . . In other words, capitalism undergoes systematic transformation only when it is forced to and when its current regime is no longer tenable. . . . *The proletariat actually invents the social and productive forms that capital will be forced to adopt in the future* (Hardt and Negri 2000, 268 [emphasis in the original]).

This constitutive ontology foregrounds the leading role of the productive forces over the relations of production. Living labor is that which "constitutes the world, by creatively modeling, ex novo, the materials that it touches" (Negri 1999, 326). It is the pure expression of individual/collective subjectivities, the cooperative power that capitalism incessantly attempts to appropriate and turn into dead labor. Biopolitics is made to coincide with the constituent *potentia* of living labor and distinguished from capitalist biopower. In this Marxian elaboration, the "politics of life" is described as an unstable patchwork of cooperative encounters.

While industrial capitalism functioned through a complex disciplinary ordering of space and time that provided the basis for the organization of the *productive cooperation* of living labor, post-Fordist capitalism no longer organizes but rather expropriates increasingly autonomous productive processes through technological networks that simultaneously connect and control. Today, Hardt and Negri write, "capitalism is increasingly external to the productive process and the generation of wealth. In other words, biopolitical labor is increasingly autonomous" (2009, 141). The paradox at the heart of global capitalism, then, is its dependence on a workforce that, although highly precarious and fragmented, is endowed with self-organizing capacities that threaten capitalist command from within. In a regime of high-tech immaterial production where the creation of value is no longer primarily attached to commodities but rather to data, code, ideas, images, sensations, and performances, capitalism functions as an apparatus of capture that aims to appropriate and valorize the innovations resulting from the field of social cooperation. In other words, for Hardt and Negri, exploitation operates primarily through the expropriation of the biopolitical common.

A central axis of Hardt and Negri's work, the common refers to many things at once. It comprises "both the product of labor and the means of future production. This common is not only the earth we share but also the languages we create, the social practices we establish, the modes of sociality that define our relationships, and so forth" (Hardt and Negri 2009, 139). The common consists of the "natural common," associated with that which is given in nature and described in terms of scarcity and limits, and the "artificial common," which includes affects, ideas, information, code, and images produced by human labor and cooperation. Unlike the natural common, the biopolitical common is described as limitless and reproducible, governed by a logic of abundance and proliferation (Hardt 2010). In this framework, the expropriation of the common under contemporary capitalism entails a double, interconnected, movement: The neoliberal policies of dispossession of natural resources and the exploitation of biopolitical labor. The latter, however, is presented as what drives capitalist accumulation and the central site of struggle for the liberation of the productive forces. The making of the autonomous common, then, is presented as the open process through which the multitude, variously described as the highly differentiated class of cognitive workers and shifting constellation of singularities, struggles for instituting a world of commonwealth beyond the constraints of public and private property. To be sure, this formulation is striking in its contrast to the Lockean conception of the earthly common as an inexhaustible repository of resources to be transformed into value by human labor and enclosures. However, a closer look at Hardt and Negri's common reveals that, despite the efforts to think of nature and society in relational terms, the former is still subordinated to the latter.

As I showed in Chapter 2, Marx had a rich engagement with the "natural" foundations of social development. Deepening this analysis, contemporary historical-materialist thinkers working on the entanglements of nature-society have developed the notions of "metabolism" and "second nature." Metabolism indicates the socio-ecological exchanges through which labor mobilizes organic and nonorganic nature in order to produce and reproduce human life (Bellamy Foster 2000; Swyngedouw 2006). "Second nature" is nature that has been transformed by human activity. The concept was used by Georg Wilhelm Friedrich Hegel to distinguish between the material environment out of history (first nature), and the fully historical complex of human institutions that manifest free will. Geographer Neil Smith contends that Marx's work straddles the line between first and second nature by offering a glimpse of the "production of nature" under capitalism. Although the refashioning of nature predates capitalism, it is only with the rise of accumulation for its own sake that nature is produced at a world scale thus achieving the "unification of all nature in the production process" (Smith 1984, 72). These approaches are insightful in that they defy the modern understanding of nature as external to society and link the transformation of nature to social relations of power. Yet, they run against the problem of grasping how "the materiality of nature" enters the process of production (Castree 1995, 20).

Hardt and Negri's conception of the biopolitical common shares a great deal with the production of nature perspective. In a conversation with Cesare Casarino, Negri claims that "mine is a nature that had experienced a process of capitalist modernization from the fourteenth century onward. In any case, the point is that for me nature is always fully cultivated nature, in which the irrigation canal is just as sacred as the tomato or the peach" (Casarino and Negri 2008, 180). Against the worship of nature as an entity out of time, Negri rightly claims that nature is not fixed and immutable but in constant transformation. This metamorphosis, however, is the result of social and cultural interactions. In other words, nature is always already "second nature," always already invested by the force of living labor. What forcefully emerges in Hardt's and Negri's work is the centrality of the productive dimensions of *bios* or living labor as "the form-giving fire" of human creative capacities (Marx 1973, 361).

At its core, Hardt and Negri's concept of the politics of life is profoundly humanist, or to be more accurate, it reflects the attempt to produce a new humanism that, while hybrid and completely embedded in the second nature of socio-technical relations, remains the principal force recreating the world. For Negri, this is the humanism emerging in Foucault's late work, a "humanism that comes after the end of any possible humanism of transcendence and that reaffirms human power as a power of the artificial, as the power to build artfully" (Casarino and Negri 2008, 146). The issue that Hardt and Negri's project of "humanism after the death of Man" (2000, 92) overlooks, however, is that of the place of a highly differentiated human in a wider field of forces that give rise to it. Saying that "nature is just another word for the common" (Hardt and Negri 2009, 171) means that, because nature is constructed and transformed through living labor, the common is the product of social practice, i.e., a social construction.

Now it becomes clearer how, for Hardt and Negri, the distinction between the natural and biopolitical common breaks down: The biopolitical common encompasses the physical world and the sociocultural practices which transform it. Still, what remains elusive is how the "mutual transformation" between human and nonhuman invoked in their ecology of the common takes place. Whereas some proponents of the "production of nature" are careful in pointing out that the recognition of the indeterminacy of nature is a crucial component in the composition of socio-natural arrangements alternative to capitalism (Swyngedouw 2011), in *Commonwealth*, Hardt and Negri turn to feminist theory to demonstrate that nature is a "subject of mutation," constantly constructed and transformed. In a passage that discusses Judith Butler's rejection of the binary of sex and gender, they write: "she argues instead that, in addition to gender, sex too is socially constructed, that sex and sexual difference are, following Foucault, discursive formations" (Hardt and Negri 2009, 170). Clearly, this is an argument against the dichotomous thinking at the heart of modernity that distinguishes between fixed biological facts and the mobile constructions of culture. Yet, by aligning themselves with Butler's insistence on the discursive matrices that limit the affirmation of nonnormatively gendered embodiments (Butler 1990; 2004), Hardt and Negri end up privileging an understanding of the common as completely *denaturalized*.¹³ What gets lost in their formulation of an ecology of the common is the "mutual transformation," the "becoming with" of nature and the social (Haraway 2008).

Whereas the autonomist analysis has for a long time set aside ecological questions as a matter "completely subordinated to industrial policies, and approachable only on the basis of a criticism of those" (Negri 2014), Hardt and Negri's more recent writing reflects the increasing preoccupation with planetary destruction. They distinguish three key terrains of expropriation of the common: the biopolitical, the extractive, and the ecosystemic (Hardt and Negri 2019) but, once again, in articulating them, they run against the limits of a political imagination and praxis that envisions a world made up of humans acting upon nature. First, in describing the biopolitical common, they incorporate feminist challenges to dualism between production and reproduction to affirm that it comprises cognitive, embodied, and affective interactions currently commodified within private property relations. Second, the account of the extractive terrain relies primarily on Verónica Gago and Sandro Mezzadra's redefinition of extractivism, a concept widely debated in Latin America in relation to the expansion of government-backed large-scale projects extracting and exporting rare minerals and fossil fuels. Gago and Mezzadra propose that scholars expand the concept of extractivism and the common to shift focus from the intensified dispossession of raw materials to broader operations of capitalism in which territories are linked to financialization. The circuits of extractivism, driven primarily by finance, target not just "inert materials" but also the networks of social cooperation, the labor and life of populations (Gago and Mezzadra 2017, 579). They connect forms of dispossession and exploitation, rural and urban contexts, sites of extraction, and global digital economies. This reworking of extractivism troubles the tendency to frame the common through the binary between natural commons and social commons. Yet, it problematically frames nature as inert, a repository of resources that are mobilized in the operations of capitalism. Third, the ecosystemic terrain considers the intimate connection between fossil extraction under capitalism and climate change. Here, again, the common is identified with the vital realms of life—the earth, the seas, and the atmosphere—that have been enclosed and depleted. The distinction between the biopolitical, the extractive, and the ecosystemic is meant to make explicit multiple aspects of the commons and to affirm that when they are brought together, they have the potential for autonomy, that is, for creating alternative social relations from within capitalism. What is missing, however, is the engagement with the imbrications across these dimensions. This has analytical and political implications. On the analytical level, while Hardt and Negri do not disavow the political relevance of the extractive and the ecosystemic terrains, their analysis foregrounds the biopolitical terrain in a way that makes it difficult to explore the nuances of the labor-nature-value nexus, that is, the shifting role of nature in the production of value (Leonardi 2017).[14] On the political level, foregrounding the biopolitical terrain over the other two implies its priority and, albeit unwittingly, suggests a hierarchy of struggles.

CONCLUSION

Written in a spirit of convivial critique, this chapter shows that in foregrounding the autonomy of living labor, Virno's, Negri's, and Hardt's theories of the common sidestep its interdependencies with the geological and biophysical milieux. In surveying the global geographies of capitalism and the terrains of struggle for the common, these scholar activists have largely overlooked ecological concerns and, until not long ago, even cast doubts about their political relevance. Although deeply materialist, they share aspects of the Anthropocene narrative of human exceptionalism that presents human collectives as the only real actors in a story of redemption. They conceive living labor as what makes the world, assembling and disassembling the earth even when seeking paths of liberation from the capitalist order. In this framework, extricating living labor from capitalism control and reorienting the trajectory of modernity is presented, particularly in Hardt and Negri's work, as a redemptive process, one leading to the emergence of a new model of the human. For Franco "Bifo" Berardi, himself part of autonomist networks, the problem is that for the authors of the *Empire* trilogy "the social potency of the common—the general intellect—is intrinsically ordained to fully deploy itself, and capitalism is intrinsically ordained to culminate in communism. But they do not consider the possibility of a stoppage in the process of deployment" (Bifo 2013, 4). Stoppages and blockages take different forms, from the psychic costs of the continuous interactions with technologies of computation, to the sheer violence of racial warfare, to new regimes of war and colonial genocide to the manifold manifestations of environmental crises ranging from the spectacular to the undetectable.

Considering its overlaps with the Anthropocene discourse, to what extent does the thought of the common contribute to producing a more just futurity? To what extent is it possible to inherit from it while at the same time departing from its tendency to ignore the entanglements between capitalist exploitation, interspecies hierarchies, and the reduction of the material world to resources?[15] One way to find answers to these questions is by looking at new scholarship and activism that have taken up this line of thinking to address pressing socio-ecological issues (Leonardi 2021). For the purposes of this book, there are three elements that remain valuable: First, the rich account of the collective that challenges liberal and neoliberal iterations of possessive individualism; second, the relentless critique of capitalism, conceived as a mutable formation always seeking new terrains of accumulation and reacting to transformations from below; third, the attunement to grassroots struggles as laboratories for thought and action. By retaining these aspects while calling into question man-then-producer as key agent of history, Chapter 4 turns to feminist intellectual and political practices that open up ways of commoning otherwise, linking together differentiated human bodies and the diverse social ecologies of which they are part of.

4

TRANSFEMINIST COMMONS: INHABITING THE EARTH OTHERWISE

In September 2020, as the first wave of COVID-19 seemed to have slowed down in Italy, feminist activists clad with face masks gathered in the garden of Lucha y Siesta, a feminist anti-violence shelter and vibrant cultural and political hub located not far from Cinecittà film studios. After months of physical distance, they assembled carefully yet joyfully for a two-day participatory design lab, the first in a series of meetings to discuss how to turn Lucha y Siesta into a transfeminist commons and thus carve out alternatives to an eviction order served by the city administration.[1]

In 2008 the Lucha y Siesta collective had reclaimed an abandoned building owned by Atac, Rome's debt-laden public transport company. Once in dismal conditions, this site has become a hub for repairing lives scarred by abuse and breathing new life into spaces neglected by owners and city officials. The two-floor house and the adjacent smaller building comprise fourteen comfortable rooms for women fleeing male violence, offices for counseling, a tailoring and recycle workshop, a library, and spaces for training anti-violence workers and hosting public events. The building is painted in pale orange with green shutters, the internal walls are decorated with feminist posters and the kitchen often welcomes visitors with enticing smells coming from simmering pots. The garden surrounding the house feels cozy, with a playground for children, leafy trees, and rich vegetation.

As a mutual-aid feminist project, Lucha y Siesta has provided a home to over two hundred women—many of them migrant and poor—and emotional, legal, and economic resources for survival to thousands more. In a city of almost three million people, where women and gender-nonconforming people fleeing abuse receive little publicly funded support, this place has been a unique outpost of feminist anti-violence practice and theorizing.[2] In addition to providing paths for repairing lives and places, Lucha y Siesta tirelessly exposes institutional failures and the systemic nature of patriarchal violence in capitalist societies built on structural inequalities of gender, race, and class (Spade 2020). This is a shelter and a political laboratory, a playground and a battleground, a working space and a resting space. Building futures through everyday practices, Lucha y Siesta's feminism centers women's collective desires rather than individualized freedoms.

In the summer 2020, however, the building was on sale. The city administrators, led by Virginia Raggi of the populist Five Stars Movement, considered it primarily as a real estate asset and a source of potential revenues for a heavily indebted city. Lucha y Siesta's experiment in creating paths of autonomy from gender violence was dismissed in the name of "legality" and the imperative of maximizing the profitability of the city's properties. Facing institutional attacks, Lucha y Siesta mounted a striking political campaign ranging from rallies and intense negotiations with municipal and regional public officials to spectacular guerrilla light projections that illuminated the city's walls and art shows featuring the work of comic artists and street artists.[3]

Bringing together various strands of political experimentations, the feminist activists turned to the commons as a grassroots project of collective management contrasting private and public property. In so doing, they built on previous movements for urban commoning that, in the midst of the European austerity policies following the financial meltdown of 2008, had reclaimed abandoned spaces and turned them into settings for practicing collective self-organization, bottom-up institutional arrangements, and the decommodification of increasingly gentrified cities (Di Feliciantonio and Aru 2018). In the case of Lucha y Siesta, the insights on urban commons has been infused both with the nuanced local perspectives acquired by working with women seeking ways out from male violence and the powerful visions developed by what the Argentinian theorist and activist Verónica Gago calls "the feminist international," the transnational movement that through massive street demonstrations, collective performances, countless assemblies, and the launch of a global strike "has made the earth shake around the world" (Gago 2020, 1).

Emerging in Argentina after the brutal murder of a young woman named Lucia Pérez in 2016, the feminist international, a movement made up by women, lesbians, trans, nonbinary, and racialized people, has found multiple localized expressions in Latin America and beyond, coalescing around the refusal of victimization and the collective desire for abolishing the interlocking violence of patriarchal rule, racism, and poverty. In a direct reference to Argentina, where women marched with the slogan "Ni una menos" (Not One Less), since 2016 in Italy the transfeminist movement has taken the name Non Una di Meno and organized through assemblies in over fifty cities.[4] Adopting the strike as protest tool, these movements have brought into sharp focus the continuum of gender-based violence against women and identified femicides as connected to broader patterns of socioeconomic and political violence. Central to the continuum of violence is the structural devaluation of the reproductive work performed by gendered and racialized bodies in contemporary capitalism. Drawing on previous feminist struggles, the strike has sought to unveil and disrupt ways in which reproductive, affective, informal, and free labor constitute an essential basis for value creation in neoliberal regimes that rely on the intensification of gender and racial injustices for governing precarity (Lorey 2015; Montanelli 2018; Gago 2020). In Italy, Lucha y Siesta has been a vital part of this movement, fueling revolt against femicide and abuse while also exploring the creation of durable feminist institutions through the commons.

During the COVID-19 outbreak, as the infrastructural breakdown of the prevalent models of distributing social and health services became strikingly visible, questions of care and reproduction acquired new salience. As many feminist and anti-racist critics pointed out, the public health crisis confirmed long-standing inequalities in modes of taking and receiving care and exposed new ones (Pirtle 2020; Care Collective 2020; Fragnito and Tola 2021; Woodly et al. 2021).[5] In the European context, it laid bare a crisis that has been long in the making with waves of privatization and financialization eroding the basis of social care (Fraser 2016; Dowling 2020). Simultaneously, the pandemic highlighted the collateral effects of the ecological destruction provoked by the economic drive for predatory land use, deforestation, intensive farming, and urbanization that create the conditions for virus spillover across species (Wallace et al. 2020). Defined by Adam Tooze (2020) as the first economic crisis of the Anthropocene, the pandemic brought into relief the intimate ties between the devaluation of care work and ecological unraveling. As such, it invited a reconsideration of reproduction to account for the socio-ecological relations in which the pandemic is embedded. The stunning spread of the SARS-CoV-2 can be seen as an expression of what Achille Mbembe defines as a planetary impasse. Conjoining long-standing historical injustices to the violence of ecological precarity, it demands nothing less than "reconstructing a habitable Earth to give all of us the breath of life" (Mbembe 2021, 62).

In opening this chapter with the assembly at Lucha Y Siesta, I draw attention to a situated project of transfeminist commoning that, acting within the Southern European context but nurtured by intense transnational feminist exchanges, openly questions the connections between heteropatriarchal violence, racism, and the neoliberal commodification of forms of life and nonlife. Questions of care and reproduction figure prominently in Lucha y Siesta's project, they intertwine with the contestation and resignification of private property through practices of commoning centered on the materiality of gendered and racialized bodies and "their relationship *within* and *with* particular places" (Pinto et al. 2014). I read the Lucha y Siesta's project as a situated mode of living otherwise that on the one hand, makes visible the continuum of gender-based violence amplified by private and public actors prioritizing profitability over life-making activities, while on the other hand invests in the process of commoning for reconstructing a more habitable city. What lessons can be learned from this experiment in transfeminist commons? How does it provide a model for reimagining the commons? Over the years, I visited Lucha y Siesta many times, participating in public events, assemblies, and workshops. My analysis draws on these visits, the examination of activist publications, as well as my participation in the larger Non Una di Meno movement in Italy.

In this chapter I consider how feminist approaches complicate the commons as a mode of production (Vercellone et al. 2017) by grappling with unresolved questions of care and reproduction that point to a more capacious sense of the commons, one that does not rely primarily on notions of labor autonomy and human exceptionalism but centers embodied placemaking practices of repairing bodies and territories. I contend that reimagining the commons through feminist practices of reproduction and care requires recalibrating these concepts to address socio-ecological dynamics. Thus, taking stock of debates in feminist theory and allied fields, I join other feminist scholars (Collard and Dempsey 2018) in proposing socio-ecological reproduction as a concept that extends feminist and anti-racist accounts of the reproductive activities that sustain life by paying attention to processes involving varied human and nonhuman beings, spaces, scales, and temporalities. If, as Silvia Federici has argued (2010), commoning is about reinventing social reproduction, then reframing this practice to account for its socio-ecological dimension carries significant implications for envisioning commons that are not limited to human cooperation. To this end, this chapter traces feminist trajectories that point toward care and reproduction as key terrains for making commons in the landscapes of the so-called Anthropocene. While Chapter 3 offered a critical assessment of the dichotomy between the social and the natural that underpins autonomist Marxist political imaginary of the common, here I weave together intellectual and activist practices that point to commons emerging from struggles against violence on bodies and places.

In this discussion I draw on feminist perspectives developed between Europe and the Americas. The interdisciplinary approaches shaping this chapter intertwine with my ongoing involvement with the transfeminist milieux in

Italy. These networks, in turn, have benefited from exchanges with movements in other localities. Latin American feminisms have been a source of collective force, opening up spaces for questioning the privileges coming with European situatedness and connecting diverse and uneven feminist geographies. My writing is indebted to this shifting feminist constellation,[6] one that brings into conversation vastly different experiences of exploitation and territorial extraction to weave distinct and yet intersecting practices of feminist and transfeminist world-making. Even as I draw concepts and desires from transnational feminist movements, it is out of the scope of this chapter to convey their uneasy complexity. My focus here is on considering the implications of a feminist reframing of social reproduction and care for the politics of the commons. I propose that such a reframing is crucial for seeking alternatives to the neoliberal exploitation of gendered and racialized bodies *and* the more-than-human worlds they inhabit. In the first part of this chapter I complicate prevalent accounts of the commons as a mode of production and stock of resources by engaging feminist perspectives that invite a reconsideration of reproduction in light of the unfolding socio-ecological crisis. This allows me to show how Italian transfeminist and queer activism have intersected and advanced these genealogies for activating modes of commoning that defy binaries of production and reproduction. My analysis consists of the close reading of activist documents circulated between 2020 and 2022, some of them as responses to the COVID-19 pandemic. Next, I consider the reworking of care and the commons in Lucha y Siesta's *Dichiarazione di Autogoverno* (*Declaration of Self-Government*), a collective text produced through a series of participatory design labs and finalized in 2022. I engage these activist materials as sources of theorizing from below that center relations between bodies and places in the mundane process of crafting commons.

REVISITING REPRODUCTION THROUGH MATERIALIST ECOFEMINISM

In recent years, amid the breakdown of social and ecological life support systems, concepts of reproduction and care have emerged as powerful counterpoints to the obsession with novelty, progress, and growth typical of modernity (Jackson 2014). They have inspired a flurry of scholarly publications, conferences, political manifestos, mutual-aid groups, and workshops (Care Collective 2020; Woodly et al. 2021; Mezzadri 2022). My work draws on feminist perspectives of reproduction and care, emphasizing their socio-ecological dimensions to analyze the commons as a process with the potential to counter and repair the consequences of "self-devouring growth," an economic model that reduces the planet's material substance to waste (Livingston 2019).

Concepts of reproduction and care are often used interchangeably but they have distinct genealogies, priorities, and commitments. The notion of reproduction has been central to the Marxist feminist critique of capitalist accumulation as a process dependent on women's unpaid labor. Care, instead, has been used in feminist political theory to challenge liberal ideas of justice and propose the rethinking of democracy through forms of relationality and interdependence (Tronto 1993). Moreover, this notion has been powerfully deployed by Black feminists as a source of collective preservation and healing in the face of deeply entrenched institutional racism (Lorde 1988).[7] While in this section, I focus primarily on reproduction as a way to complicate Marxist accounts of the commons as a mode of production, I return to care later in this chapter.

The concept of social reproduction was first used by the eighteenth-century economist François Quesnay in the *Tableau économique* to indicate the process by which a social system reproduces itself. From the perspective of classical political economy, the key concern was how individuals and classes making up a complex society reappear after each productive cycle. Since the 1970s, reproduction has been critically reconfigured in feminist theory and activism as a field of struggle between divergent modes of organizing the activities that make and regenerate life, including domestic work, sex, and the care of the elderly (Federici 1975).

The concept has been recently revamped to account for the current crisis of care and the gendered and racialized divisions of reproductive labor in global capitalism (Fraser 2016; Battacharya 2017; Mezzadri 2019). While much attention has been paid to the reconfiguration and commodification of reproduction in neoliberal times, the socio-ecological dimension of reproduction has remained less explored. In contributing to rework this notion, I turn to a particular genealogy of the concept, one that explores the encounter between 1970s feminist critiques of reproduction and materialist ecofeminism.[8] Examining Silvia Federici and Mariarosa Dalla Costa's work in conjunction with contributions by Maria Mies, Vandana Shiva, Ariel Salleh, and other feminist scholars, I interrogate the nexus between the appropriation of nature as resources and the exploitation of gendered and racialized reproductive labor across uneven capitalist geographies. I explore reproduction as a shifting assemblage of more-than-human life and nonlife, that, far from being confined to women's labor, encompasses a wider set of activities and processes. These include ongoing enclosure, organized dispossession and expansive extractive operations involving humans and nonhumans in "essential yet under-recognized shadow realms of expropriation" (Lowe 2020, 223). This approach to reproduction has implications for theorizing the commons beyond the focus on human labor power, as a terrain of appropriation, exploitation, and resistance that encompasses gendered and racialized bodies as well as the ecologies of life and nonlife that sustain them.

In the 1970s, the campaign Wages for Housework demanded remuneration for the unpaid activities that women perform within the family in capitalist societies. In Italy and the United States, scholar activists such as Mariarosa Dalla Costa, Selma James, and Silvia Federici contended that under patriarchal capitalist relations, reproductive work is devalued, naturalized as a feminine vocation, and freely appropriated. Drawing on Marx and yet challenging the Marxist prioritization of production over reproduction and the overwhelming focus on the male worker as the agent of revolutionary transformation, they maintained that insofar as care work reproduces and sustains labor power, it contributes to value creation (Dalla Costa and James 1975; Federici 1975). Wages for Housework identified social reproduction as a site of struggle, one that needs to be radically reconfigured in order to create modes of living beyond the grip of capitalism. In critical conversation with Wages for Housework, Black feminists noted that reproductive labor is stratified along racial and gender lines and informed by hierarchies established through colonial plantation economies (Davis 1981). In doing so, they opened up space for contemporary analyses of gendered and racialized work in global care chains (Boris and Parrenas 2010) as well as the transnational politics of assisted reproductive technologies in contemporary biocapitalism (Vora 2015).

In the decades since the 1970s, the concept of social reproduction has remained central to the work of Marxist feminists such as Federici and Dalla Costa while also shifting to integrate ecofeminist perspectives.[9] Materialist ecofeminism, particularly the work of the German sociologist Maria Mies and the Australian scholar Ariel Salleh, has provided crucial input for connecting the devaluation of reproduction to the modern European project of mastering nature. Writing in the 1980s, Mies builds on Rosa Luxemburg's analysis of capitalism as depending on an "outside," that is, uncommodified forms of living to draw on for the continuous process of accumulation (Mies 1986). In Mies's account, ongoing accumulation deploys violence as the means for the simultaneous exploitation of women, nature, and colonies. Tying together distinct forms of exploitation, Mies traces a web of "underground connections" linking the subordination of women in Europe and the dispossession and dehumanization of colonized subjects (ibid., 77). Extending Carolyn Merchant's (1980) analysis on the "death of nature," Mies contends that the modern European notion of progress, and later that of development, has been predicated upon the separation of man from nature, treated as a reservoir of resources to be controlled through technological tools. At the same time that nature is turned into inert matter, women have been separated from the means of subsistence and forced into the devalued, naturalized, realm of reproduction and housework. Mies considers European colonialism as a foundational process for European "progress" and "development." Within modernity, the colonies and the colonized have become "defined into nature" with the plundering of land and the extraction of labor from slaves and Native populations. The progress of European men, she concludes, has depended on unpaid women's work in the household, the exploitation of nature's work and that of colonized subjects. Over the course of five centuries, these have been the reservoirs of exploitation to which Marxism has largely remained blind. Like the underwater part of an iceberg, they have formed the invisible basis of the capitalist world economy, that which has made its reproduction possible. While "capital and waged labor form the visible economy, 'above the water,' counted in the GDP, where waged labour is protected by a labor contract, and where housework, work in the informal sector, work in the colonies and nature's production form the underwater part of this economy" (Mies 1986, xi). From this perspective, reproduction indexes a web connecting human activities with natural processes.

This point has been further articulated by Ariel Salleh through the notion of meta-industrial work, defined as the complex interaction with the material world directed at maintaining "the necessary biological infrastructure for all economic systems" (Salleh 2004, 9). Meta-industrial work, another name for reproductive labor, is the node point for a range of activities that "enhance metabolic flows between humans and nature, and flows within ecosystems" (ibid., 9). These include activities that are at the margins of capitalist production and yet essential for its reproduction such as unpaid or underpaid work of caring for children and the elderly, subsistence farming and gathering, urban gardening, and all forms of maintenance and repair involving humans and nonhumans on earth. Unlike the extractive capitalist mode of production oriented toward manufacturing commodities and the externalization of costs, meta-industrial work creates "metabolic-value" in that it engages in processes of socio-ecological regeneration. Because this value has been externalized and freely appropriated within the dominant regime of production, capitalism owns debt on three levels: 1) social debt to low-paid workers; 2) embodied debt for reproductive labor; and 3) ecological debt for ecosystem degradation. The triangulation of these three elements, for Salleh, has political implications in that it points to alliances among diverse meta-industrial workers.

Since at least the early 1990s, encounters with materialist ecofeminism and a range of movements resisting environmental violence have pushed Dalla Costa and Federici to revisit the notion of reproduction. Moving away from the exclusive focus on housework, their critiques of the capitalist appropriation of so-called externalities to the market have connected the plundering of the earth with gendered and racialized forms of dispossession. Dalla Costa (2005) has described her journey from the 1970s militancy with the Marxist group Potere Operaio and Wages for Housework to the more recent research on economies of subsistence and the environmentalism of the poor as a journey from the kitchen to the land. She has recalled how the kitchen, once a space of exploitation for many women within the "social factory," has become a site of counterplanning. Similarly, the land, at once a terrain of expropriation and life-making activity, has been a site of struggle but also a source of sensation and imagination. The politicization of reproduction cuts across these different spaces, shifting from the focus on time and money through the demand of a wage for domestic work, to the reproductive powers of land threatened by privatization, toxicity, and widespread ecological destruction. As Dalla Costa (2019) explains, "The question of the land overwhelmingly forced us to rethink that of reproduction" (ibid., 242). This shift has occurred through exchanges and activist research with networks for food sovereignty, Indigenous Zapatistas from Chiapas, mothers from Plaza de Mayo, small peasant and consumer organizations in Italy, and ecofeminists including Mies and Vandana Shiva. From the engagement with the politics of food production and consumption came the understanding of the earth as that which enhances human capacities for feeling, thinking, and acting together. Land, writes Dalla Costa, "is not only our source of nourishment, but it is from the land that bodies gather meaning, sensations, and imagination" (ibid., 231).

In *Our Mother Ocean* (2014), a book written with Monica Chilese, Dalla Costa broadens the scope of reproduction to include the consideration of the ecological exhaustion and toxicity provoked by the expansive neoliberal enclosures of the seas. The book documents the devastating effects of the globalization of the fishing industry and how the World Fisher Movement that first emerged in southern India has struggled to protect the ocean as source of food, livelihood, and attachments for human and nonhuman life forms. Dalla Costa defines the ocean as a commons whose regenerative powers must be supported against neoliberal privatization through the cooperation between humans and nature. This position echoes Carolyn Merchant's call for a "partnership ethics" (2003) grounded in relationality and reciprocity that understands humans and nonhuman nature as equal partners able to cooperate in the achievement of mutual survival. It also recalls Shiva's image of the "Earth family" (2006), a reference to the community of beings supported by Mother Earth. For Shiva, as well as Dalla Costa, the care of the commons, the earth that we share with nonhuman living beings, would require attention and appreciation for the life-sustaining practices performed at the level of the community, particularly by women and Indigenous groups in the Southern Hemisphere. The making of the commons, in other words, would proceed hand in hand with a re-enchantment of the world, a consideration of nature as harmonious whole, capable of regeneration and dynamic equilibrium.

Federici shares a similar trajectory, one moving from the focus during the 1970s on the exploitation of women in the process of capitalist accumulation to a broader analysis of the neoliberal order predicated upon economic policies that target gendered and racialized bodies and their relations to land. Her account of reproduction has been shaped by her time spent in the United States and Nigeria, where she developed connections with a range of activist groups and social movements (Federici 2020). In the United States, Federici collaborated with Midnight Notes, a collective formed after the Three Mile Island nuclear accident in 1979. Over the years, the collective developed heterodox Marxist analyses of the work/energy crisis, anti-nuclear movements, and globalization, while at the same time forging strong connections with anti-colonial activists and intellectuals in Mexico and Guatemala. While living in Nigeria in the 1980s, Federici had ties with local feminist organizations facing the effects of the World Bank's structural adjustments programs that deprived them of access to land, water, and means of subsistence. These encounters and continuous conversations with materialist ecofeminists and, more recently, with feminist movements in Latin America, have led Federici to reconsider reproduction far beyond "women's labor" and unpaid housework. Her revised account of reproduction reflects a concern for the links between the restructuring of reproductive work through global care chains, the ongoing legacies of colonialism and slavery, and the ecological costs of extractivist economies relying on land privatization and nature commodification. As Federici states, "any theory of social reproduction should also be concerned with the ecological struggle and the struggles of Indigenous people. These are powerful social movements that cannot be ignored, and that demand we rethink Marx's work in some important respects" (Federici 2020, 157).

In attending to the socio-ecological costs of capitalist development, Federici explicitly diverges from the influential autonomist tradition associated with the work of Michael Hardt and Antonio Negri (see Chapter 3). Post-workerist theorists tend to focus their attention on the highest levels of capitalist development, often overlooking struggles happening at the margins and the peripheries of the global economy. In doing so, they risk privileging a particular set of workers, namely the cognitive workers of the creative industries and digital economies while failing to notice that "online communication/production depends on economic activities—mining, microchip and rare earth production—that, as presently organized, are extremely destructive, socially and ecologically" (Federici 2010, 286–87). This critique is important for the purposes of this book: While the autonomist Marxist approach foregrounds the common as cooperative human labor, feminist critiques of reproduction highlight the socio-ecological conditions that make this specific mode of human relationality possible.

Faced with the continuous recreation of gendered and racialized hierarchies and increasing environmental violence within the globalized economy, Federici theorizes the commons as a form of collective reproduction outside of the logic of the market and the state. Her feminist accounts of the commons, which she developed starting in the early 2000s through observation and participation in social struggles, highlight three key elements. First, she maintains that commoning is a conflictual practice, one that emerges in antagonistic relationship to the state and the market. In this sense, she cautions against the ways in which the commons can be adapted to function within market-based economies, for instance through the top-down creation of "global commons" that deprive local communities of access to land. Second, in line with Mies and Veronika Bennholdt-Thomsen's subsistence perspective (1999), feminist commoning requires the reclaiming of collective modes of living based on the redistribution of the reproductive activities historically performed by women, including domestic work. Commoning means reclaiming common wealth through struggles against the privatization of social services but also for access to the means of subsistence, land, water, and sources of food production. Third, for Federici (2019) the commons are not resources but a set of "re-enchanted" eco-social relations that involve human beings, animals, plants, and minerals. Through commoning, she argues, reproductive activities can turn from a source of exploitation to one of experimentation. What was once refused as stifling and isolating gendered activity can be valorized through the collective engagement with nonexploitative modes of living involving human and nonhuman beings.

The feminist scholars discussed in this section have not only reworked the category of labor to make visible historically devalued human activities, but they have shifted the meaning of reproduction and care to account for the appropriation and exploitation of land, minerals, water, plants, and animals. To be sure, this body of work is not devoid of problems. As my reading of Dalla Costa makes clear, at times she renders nature as a homeostatic system with which it is possible to cultivate a partnership ethic of earthcare based on a relation of mutual dependence and the ability to cooperate on equal ground. Yet this view is difficult to sustain in light of the positive feedback loops and irreversible "tipping points" that characterize complex earth systems.[10] Moreover, the alignment between atmospheric and oceanic forces and the universalized figure of the nurturing mother has been challenged in feminist and queer environmental scholarship and activism (Tola 2018). Not only does it risk reproducing the conflation between women, motherhood, and care work, but it also risks eliding the question of alterity. For, as Gayatri Spivak points out in distinguishing between the global and the planetary, "the planet is in the species of alterity, belonging to another system; and yet we inhabit it, on loan" (Spivak 2003, 72). The earth is neither a nurturing mother nor a partner. It is an ensemble of concatenated differences, the medium connecting different scales, from planetary immensity to the soil on which one walks (De la Bellacasa 2017). In Frédréric Neyrat's (2019) words, the earth is "a form of alterity, an outside," a historical trajectory lasting a billion years and carrying within itself uncountable forms of life and nonlife (ibid., 65). Adding to this, Federici's insistence on the nexus between the commons and the community, understood as the site for reclaiming access to the land, risks reproducing the vision of the community as a harmonious unit. Although she has often described the commons as a process, her revaluation of care work within the community at times evokes a holistic whole, unruptured by power struggles and asymmetries.

These disagreements notwithstanding, my point here is that these materialist ecofeminist perspectives provide generative insights to illuminate socio-ecological reproduction. They contribute to foregrounding how "the work of nonhuman nature is placed in a structurally similar position to other reproductive work" (Collard and Dempsey 2018, 1356). Shifting focus on socio-ecological reproduction helps us to consider the connections linking the exploitation of gendered and racialized work to ecological dispossession in varied landscapes of capitalism. Importantly for the purposes of this book, this concept poses serious challenges to the Marxist visions of the common as grounded in species-specific capacities and the potential autonomy of living labor in current technologically supercharged modes of production. Whereas, as I show in Chapter 3, autonomist Marxism tends to foreground the production of man by man, and the common by human beings, socio-ecological reproduction shifts attention from the teleology of labor to the milieu of living and nonliving forces on which human beings depend. In this book I use socio-ecological reproduction for reimagining commoning as the practice of decommodifying the relational networks through which more-than-human collectives reproduce themselves and assert their autonomy from capitalism. This entails considering how the interdependencies among gendered, racialized, and classed bodies and more-than-human ecologies play out in the process of making commons and how they create alternatives to capitalist processes of exploitation and commodification of living and nonliving beings. In my work, socio-ecological relations rather than just human living labor are at the heart of the politics of the commons.

Drawing on materialist ecofeminism, recent research has addressed the multiple valences of reproduction in contemporary capitalism. Jason Moore (2015) has coined the term "cheap natures" to define the appropriation of unpaid or low paid forms of work/energy in the history of the Capitalocene, his preferred name for a particular way of organizing nature that emerged in Europe in the late sixteenth century. With a more decidedly feminist outlook, Stefania Barca (2020) builds on materialist ecofeminism, particularly the work of Mies and Salleh, to direct attention to the "forces of reproduction," that is, the feminized, racialized, and dispossessed subjects whose meta-industrial work keeps the world alive by taking care of the biophysical environment. The term "forces of reproduction" identifies human and nonhuman reproductive activities as terrains of devaluation and exploitation. At the same time, it indexes a political subject made by the convergence of a range of life-making struggles that value the political relevance of nonhuman forces. Other scholars have drawn on Marxist feminism for analyzing the ways in which biological and ecological processes have become enmeshed in the circuits of capitalist accumulation. In the neoliberal context, reproduction has taken on new relevance as site of value creation. Within contemporary bioeconomies that straddle the divide between society and nature, human and nonhuman regenerative processes have been enrolled as "hybrid labor" in the transnational circuits of capitalist accumulation (Battistoni 2017; Cooper and Waldby 2014). In the same way that a range of (re)productive activities previously invisible have become commodities, with the unfolding of climate change the risks associated to environmental complexity have entered the sphere of economic calculus.[11] From disaster management to low-carbon markets to ecosystem services and weather derivatives, contemporary forms of value extraction thrive on instability and uncertainty (Cooper 2010; Pellizzoni 2025). These interventions make clear that human activities, lives, and spaces are not the only sites of commodification.

Extending theorizations of socio-ecological reproduction, this chapter contributes to feminist scholarship that reframes labor by considering the conjoined commodification of human and nonhuman activities in the circuits of expanded extractivism (Gago and Mezzadra 2017). I propose to define socio-ecological reproduction as a field of struggle between divergent ways of configuring the regenerative activities of gendered and racialized human beings in relation to the living and nonliving ecologies that make human existence possible. What would it mean to reinvent socio-ecological reproduction in ways that challenge and create alternatives to its current capitalist configuration?

How can the reinvention of socio-ecological reproduction contribute to the necessary work of reparation to make earth habitable again? There is no one answer to these questions, but there are multiple avenues and scales for exploring and experimenting. As a way to contribute to this endeavor, in the next section I think with activist collectives that point toward the commoning of socio-ecological reproduction. While my primary focus is on Italian transfeminist and queer networks, I include references to transnational feminist engagements with the nexus between gender and racial violence, extractivist economies, and ecological destruction. Rather than reflecting on reproduction and care as human affairs, these activist interventions attend to multiscalar connections bringing together economies and ecologies and involving multispecies relations and varied temporalities.

ACTIVIST REFRAMINGS OF REPRODUCTION

Since the early months of the COVID-19 pandemic in winter 2020, feminist critics highlighted how the management of the outbreak demonstrated the continuous devaluation of reproductive activities that make and sustain life, including health care, education, and domestic work. Black and anti-racist feminist scholars and activists directed attention to stark care inequalities along the intersecting axes of gender, race, class, age, and geography, for instance illuminating the disproportionate impact of SARS-CoV-2 on Black and Brown communities in the United States (Mezzadri 2022; Pirtle 2020). In Italy and the United Kingdom, governments often praised hospital and other "essential" workers as heroes on the front line of the battle against the virus, yet they continued to privilege production over reproduction. Italy was the first Western country to be heavily hit by COVID-19. Even as the contagion was surging in late February, the local government gave in to the pressures of industry organizations that insisted on keeping production going in nonessential factories and workplaces (Rispoli and Tola 2020). In the United Kingdom, Boris Johnson insisted on getting "Britain back to work" for months before being forced to impose new rounds of national lockdown to flatten the curve of contagion. Across the global political spectrum, governmental narratives figured the impact of COVID-19 mainly in economic terms, through the loss of productivity and jobs. Complicating this view, feminist and anti-racist scholars maintained that this public health crisis has been part of a larger crisis of reproduction, reflecting distinctions between disposable lives and protected lives in neoliberal regimes that have been dismantling welfare provisions and individualizing care responsibilities. In this context, networks of mutual aid and solidarity have proliferated, often invoking the commons as a means of sustaining life (Care Collective 2020; Ticktin 2021).

While these critical perspectives on care and reproduction are vital, they need to be supplemented with considerations about the wildly uneven socio-ecological transformations that underpinned the pandemic outbreak and, more broadly, climate change and other environmental disasters caused by the resource logic of colonial capitalism. A zoonotic disease, COVID-19 spread across the circuits of globalization, its effects playing out in localities shaped by specific histories and policies, including ongoing legacies of slavery and colonialism, the neoliberalization of health care infrastructures, and political decisions that render human groups differentially vulnerable to outbreaks. As an event occurring in particular places with global and yet localized effects, COVID-19 is one of the many unintended consequences of capitalist ruination (Tsing 2015). Its emergence and lethality depend on uneven socio-ecological relations of appropriation as well as heterogeneous forms of exploitation and disposability.

Italy went into strict lockdown the night of March 7, 2020, forcing the local feminist movement to reconfigure the mobilizations planned in the following days in conjunction with the International Feminist Strike. Breaking social isolation, activist networks launched a variety of radical care initiatives (Hobart and Kneese 2020). Domestic violence shelters such as Lucha y Siesta continued to operate remotely, online services supported women living with abusive partners and those seeking access to abortion, and sex workers collectives launched crowdfunding projects to raise money for those most hit by the lockdown. These efforts joined forces with broader grassroots initiatives to support the most vulnerable groups. First in Milan, then in other cities, solidarity brigades mobilized to help people in need with home delivery of groceries and medicines. Radical unions opened helplines to provide legal support to workers facing safety issues and job losses. Meanwhile, strikes in factories, warehouses, and ports signaled workers' refusal to choose between health and livelihood and riots erupted in overcrowded jails. As activist networks mobilized, they widely deployed feminist interpretations of reproduction and care to account for the socio-ecological aspects of the pandemic crisis.

In late April, the feminist assembly Non Una di Meno Rome, of which I was part at the time, circulated *Life Beyond the Pandemic*, a collective essay mapping the experience of the lockdown through the lenses of social reproduction.[12] This text is an attempt to make sense of the pandemic as part of a larger reality of precarity and violence under neoliberal capitalism and come up with collective strategies of resistance. It considers reproduction as a field of inequality in which the exploitation of gendered and racialized bodies performing "essential" work cannot be separated from the dismantling of social services, the legacies of colonialism, racism faced by migrants, and the intensification of large-scale environmental exploitation that has created the conditions for the emergence of the pandemic. Considering the catastrophic response of Italy's health care system in the wealthiest part of the country, the essay argues that the "the collapse of healthcare follows that of ecosystems" (Non Una di Meno Roma 2020, 112). This means that as the coronavirus traveled from devastated wild ecologies to human bodies, moving across borders, its lethal action in the Italian context was amplified by the dismantling of public health care systems through over a decade of austerity measures. This feminist account of the specificities of the Italian care crisis considers reproduction as encompassing social and ecological processes, with human and nonhuman actors differentially caught in the expansive dynamic intrinsic to capitalist growth.

Articulating feminist debates with insights from ecological movements, the essay continues by arguing that, in light of the pandemic "the connections between social and ecological reproduction can no longer be ignored or dismissed as secondary" (Non Una di Meno Roma 2020, 113). This requires bringing feminist insights on the redistribution of care outside the nuclear family to bear on the ecological crisis. It demands redefining reproduction to account for the life-making activities emerging from the complex interactions linking humans to nonhuman others. In political terms, such a reframing of reproduction has the potential to open up spaces of alliance: "it is the ground of encounter, and possible convergence, between feminist, transfeminist and ecological movements" (ibid., 113).

Along similar lines, a blog post written in late April 2020 by B-side Pride, a queer network based in Bologna, points out that the pandemic "has shown the vulnerability of all bodies, and their interdependence with other species, populations and territories. It has clearly demonstrated the limits and contradictions of the social and economic model that many call 'normal.'" [13] Here, questions of bodily vulnerability and interdependence are linked to territories, somehow echoing the praxis of *cuerpo-territorio* (body-territory) developed in Latin America by rural Indigenous women mobilizing in defense of land in the context of extractivist economies (Colectivo Miradas Criticas 2017).[14] B-side Pride notes the intensification of precarity experienced by queer bodies during COVID-19 while at the same time extending notions of queer kinship to other-than-human beings. Queer and trans lives, particularly racialized ones, have been rendered more precarious by forms of viral containment centered on the confinement of individuals within often inhospitable family homes. Queer activists disrupt the notion of the heteronormative nuclear family as a protective space by highlighting the role of communities of mutual support that exist outside the traditional domestic sphere. In so doing, they also put pressure on feminist notions of care and reproduction focusing on "women's domestic labor within heteronormative households" (Malatino 2020). In the same vein as Hil Malatino's analysis of trans care, B-side Pride draws attention to an alternative set of locations and relations, to modes of kinship formation that are not constrained within the idealized White middle-class nuclear family but occur through queer and trans networks of intimacy. Still more, the queer kinship evoked by the Italian activists gestures toward a field of interspecies dependencies that connects specific locations to processes of socio-ecological transformation at the planetary scale. As the product of an unsustainable economic model that thrives on the continuous encroachment on new spaces of life, the pandemic requires situated responses to complex dynamics that bring together seemingly unconnected species, populations, and territories.

B-side Pride's post concludes: "what we need is not to hope to go back to normal, which for us was the problem. Rather, we need to rethink the basis of social and ecological re/production" (2020). Building on a widely circulated slogan from Chile, the text expresses the refusal of returning to the normal state of affairs and argues for the reinvention of reproduction to do away with heteropatriarchal and capitalist paradigms that protect the lives of some while systematically intensifying conditions of socio-ecological precarity for many. Following up on these reflections, the queer network launched a citywide mutual aid initiative that provided material and emotional support to queer and trans people, particularly refugees and asylum seekers. As part of these efforts, it distributed refurbished laptops to help those at risk of isolation stay connected, and collaborated with Campi Aperti, a local food project oriented toward the commoning of food production and consumption (De Angelis and Diesner 2020). To support access to this alternative food system, the queer collective raised funds that were converted into "Grani" (the Italian word for grains), a special currency for purchasing fruit and vegetables at Campi Aperti. This initiative combines the queering of care with the need to shorten the supply chain through initiatives for developing food sovereignty. In this sense, it experiments with the commoning of socio-ecological reproduction. These intellectual and political practices of feminist and queer activists adapt the concept of reproduction for a time of overlapping and uneven crises (Sultana 2021). Amid the intensification of gender, racial, and economic violence exacerbated by the pandemic, they identify the systemic nature of socio-ecological inequalities while also pointing to the commons as a pathway for undoing them.

COMMONING CARE

The political campaign to reclaim Lucha y Siesta as a transfeminist commons has involved a multigenerational multitude of anti-violence activists, women living in the shelter, artists, lawyers, and academics. Many of the women sheltered in the house refused to be frozen in the role of victims whose life is managed through inadequate institutional forms of care and became powerful voices of contestation and transformation, fighting at once for their lives and for a collective project. The prolonged struggle led to a stunning achievement: The regional administration, yielding to intense political pressure, decided to purchase the building and announced the intention to formally recognize its status as a commons managed by the Lucha y Siesta collective through a participatory process. This development, however, was short-lived. At the time of this writing, in June 2024, the right-wing governor of Lazio, the region around Rome, withdrew from the agreement negotiated under the previous center-left governor, thus forcing the activists to shift back into campaign mode to defend the place.

The *Dichiarazione di Autogoverno (Declaration of Self-Government)* finalized in June 2022, has emerged out of Lucha y Siesta's process of commoning.[15] The document comprises a series of sections introducing the project of feminist commoning, its guiding principles, objectives, and protocols of organization. Written in an evocative yet accessible style, it foregrounds the power of the collective, a "we" that acknowledges and builds on the experiences of women, lesbians, trans, and nonbinary people. Based on my observation over the years, Lucha Y Siesta remains a space mainly managed by White cisgender women. Yet, it has developed a keen awareness of the multiple oppressions faced by bodies differentially marked by gender, race, class, and citizenship status. This, among other things, has translated into initiatives that problematize Whiteness within feminist movements and center the experiences and perspectives of Black Italians, migrant women, and other racialized bodies. The *Declaration* thus reflects attention to intersectional histories, the dynamics of race and racism in postcolonial Italy, and the difficulties in composing alliances across different experiences of power and privilege.[16]

Overall, Lucha y Siesta's objective is contrasting gender-based violence. But violence here, as in the Non Una di Meno movement, takes a complex meaning as the product of deep-seated and interconnected gender, racial, economic, and environmental inequalities. The *Declaration* describes transfeminist commoning as the prefigurative process of creating alternatives to violence.[17] This means that this project is "defined around practices and knowledge that are constantly in flux," situated in the "here and now and simultaneously . . . oriented to transforming the present state of things" (Lucha y Siesta 2022, 3). In other words, transfeminist commoning embodies ways of living otherwise in particular times and spaces, while at the same time envisioning broader alternative futurities displacing the primacy of private property and individualized freedoms for a privileged few.

Concept and practices of care are central to this transfeminist commons. Care refers to "the huge amount of invisible and unpaid (or severely underpaid) care and reproductive work done within family and social relationships" (Lucha y Siesta 2022, 5). At the same time it points to the embodied and affective activities "of maintaining spaces and relationships . . ., as the vital political practice in interdependent worlds with ethical and affective implications" (ibid., 17). These passages resonate with feminist and anti-racist genealogies that present care as a deeply ambivalent practice (Fragnito and Tola 2021). Care indexes a devalued set of activities, historically excluded from the realms of politics and economics (Tronto 1993). While some subjects, often White men, have been constructed as deserving care, others, usually gendered and racialized subjects, have been burdened with the task of providing it through a range of everyday activities, including cleaning, picking up trash, tending for children and the elderly, and providing food and comfort (Boris and Parrenas 2010; Raghuram 2021). The labor of caring for others as well as the possibility of accessing care have been unevenly distributed thus creating acute asymmetries and exclusions along lines of gender, race, class, and geography. Lucha Y Siesta's modes of contrasting male and heteropatriarchal violence from below show that care comes in diverse configurations. While some of these have been coterminous with patriarchal, racial, and capitalist logics, others work to undo hegemonic power formations even as they remain implicated in them. While top-down institutional programs tend to address violence through the discontinuous and chronically underfunded provision of care for individualized victims, transfeminist practices turn care into a terrain of contestation and collective processes of subjectivation made up of everyday activities. In order for transfeminist care to emerge, hegemonic care arrangements have to be called into question and undone.

These passages of Lucha y Siesta's *Declaration* seem to align with Joan Tronto's oft-cited formulation of care as "a species activity that includes everything we do to maintain, continue, and repair our 'world' so that we can live in it as well as possible. That world includes our bodies, ourselves, and our environment, all of which we seek to interweave in a complex, life-sustaining web" (Tronto 1993, 103). But I suggest that Lucha y Siesta's transfeminist commons extends Tronto's approach to care in two ways. First, over the years, the collective has performed an invaluable work caring *for* and *with* women and children experiencing high degrees of psychological, physical, and sexual abuse. This form of radical care (Hobart and Kneese 2020) has allowed many to live through hardship in a context of institutional neglect and the increasing privatization of health and social services. Still more, it has turned a forgotten urban site into a flourishing place. Recalling these practices, the activists have claimed the building as a form of partial restitution for "an incalculable care debt" owned by local institutions (Lucha y Siesta 2022, 5). Thus, they complicate and extend Tronto's theory of caring democracy by contesting privatization and affirming practices of collective and inclusive use as a way to challenge at once the lack of institutional care and the top-down approaches that often characterize institutional care settings. Second, they provide a distinctive account of the relation between bodies, spaces, and places. Let me elaborate on the last point. Lucha y Siesta identifies as "a dispositive of care and self-care involving bodies that are always in relation with space because neither bodies nor space are neutral and fixed" (ibid., 7). Further, it adopts an intersectional and ecological posture "in relation to space, to the surrounding territory, and to other human and nonhuman bodies" (ibid., 7). This suggests that transfeminist commons enact a practice of inhabiting otherwise. As discussed in Chapter 2, prevalent modern European forms of habitation focused on making space productive through appropriation and enclosures leading to the improvement of land and individuals.[18] In the current context, as spaces of power relations rather than mere containers or support for human action, European cities have become central hubs in the government of the precarious, that is, the instrumentalization of precarity to render stratified populations governable (Lorey 2015). As the urban geographers Ugo Rossi and Arturo di Bella note, the European urban spaces of late neoliberalism have been laboratories for the overlapping processes of dispossession through new forms of valorization centered on digital platforms that organize the performances of entrepreneurial selves (Rossi and Di Bella 2017). Lucha y Siesta's transfeminist commons deploy feminist solidarity to address the structural forms of gendered and racialized violence produced by the dense overlap of dispossession and valorization in urban spaces. In contrast with the neoliberal injunction to inhabit spaces as entrepreneurs of the selves, the transfeminist commons gain strength from collective and embodied forms of relation with space and place.[19]

Drawing on Mathis Stock's notion of habitation as "making with space," the environmental humanities scholar Stephanie Posthumus (2017) describes ecological dwelling as a practice grounded "in the physical processes, exchanges, and interactions of bodies in place" (ibid., 61). Practices of dwelling, she writes, "are ecological when they take into account the materiality of the space being made" (ibid., 61). The transfeminist commons considered here foreground the materiality of bodies and spaces as they make each other in a process of placemaking that contests modes of urban habitation revolving around private property, the individualized freedoms of the entrepreneurial self, and its reproduction within the heteronormative nuclear family. Interweaving mundane activities that repair bodies and create radically convivial places, Lucha y Siesta reinvents reproduction and care and enacts conflicts for reclaiming the care debt which is owed to those whose labor has been keeping the world alive (Barca 2020). In so doing, they contribute to making the city, and the earth, habitable again.

CONCLUSION

My work in this chapter has been to thread together intellectual insights on the commons with activist practices in Italy to illuminate feminist, transfeminist, and queer commons as alternative modes of inhabiting European cities that are becoming increasingly inhospitable for precarious workers, dissident subjects, migrants, and the poor. By way of closing, I want to return to what transfeminist commons can offer in times of overlapping crises and "co-producing injustices" (Sultana 2021).

First, I address the relation between commons, public, and private actors in neoliberal European contexts. Some would characterize the feminist, transfeminist, and queer projects of commoning discussed in this chapter as merely filling the gaps created by the state withdrawal of welfare provisions and thus unwittingly responding to the neoliberal injunction to become fully responsible citizens capable of reorganizing services that the state refuses to provide. However, these activist initiatives do not prioritize individual or community resilience in the face of state abandonment. On the contrary, they identify and challenge the power structures that have created the conditions for the crisis and attempt to craft ways out of it. For instance, in *Life Beyond the Pandemic* Non Una di Meno Roma makes clear that "struggles for reconfiguring the infrastructures of care" are directed at "taking control away from market forces" (2020, 113) and are intertwined with demands for expanding welfare services to make them accessible to all. Rather than placing the private and the public on the same plane and opposing the commons to both, the essay argues that "to defend the public means to imagine common institutions and freedom beyond the State" (ibid., 113). This quote implies that the commons are institutions from below that challenge the dominant state form while entering into complex relations with its political structures. Lucha y Siesta's *Declaration* points out that the goal of transfeminist commons is not to "provide services"

(2022, 7). It does not replace the institutional actors that are charged with the responsibility of delivering them but it demands institutional accountability while also experimenting with alternative relations between service providers and service receivers where the latter are not assumed as passive actors in a top-down process.

Second, I turn to the multiscalar reach of feminist, transfeminist, and queer commons that encompass bodies, places, and environments. The intellectual and activist contributions presented in this chapter revisit notions of reproduction and care by connecting everyday life-making activities that repair specific bodies and places to broader struggles against escalating racial, gendered, and environmental violence within neoliberal regimes. They theorize and enact the commoning of care and repair, foregrounding the political relevance of practices that involve gendered and racialized subjects in relation to places and environments. Put another way, feminist, transfeminist, and queer commons create space for a form of caring "for different kinds of histories at the same time and find ways to keep them in the same frame, despite their different scales" (Ticktin 2019, 151). Situated in a Southern European space of state divestment from social reproduction and extractive capitalist dynamic, these projects highlight forms of care rooted in place but capable of contesting modes of accumulation that have rendered the planet inhabitable. In political terms, these practices translate into seeking alliances across movements striving to revalue human and nonhuman regenerative processes. Such efforts aim to undo dominant regimes of valuation and to redistribute the very possibilities of life on earth. Attending to the multiple dimensions of reproduction and care, transfeminist commons contribute to hold open the space for inhabiting the earth otherwise.

PART III

POLITICAL ECOLOGIES

5

COSMOPOLITICAL COMMONING IN A CITY OF RUINS

In the late-afternoon light the area recalls the drowned worlds imagined by science fiction writer James Ballard (1965). Decaying industrial buildings are reclaimed by resurgent vegetation. Next to the ruins, thick brambles, rushes, and willows flank an urban lake. The skeleton of a parking lot is half submerged in glimmering water, peopled by a myriad of birds, insects, and reptiles. This uncanny landscape is part of the Ex-SNIA Viscosa, a former chemical-textile complex in the densely populated Prenestino-Labicano neighborhood in Rome, Italy. In this chapter I explore the Ex-SNIA as a site of commoning. I am interested in how the urban lake became a major actor, rather than a "resource," in a political project that has been challenging processes of gentrification and creating modes of living otherwise in a city of uncontrolled expansion, inadequate welfare provisions, swelling public debt, and increasing precarity. At the Ex-SNIA, the force of the lake is part of a field of struggles connecting the past and the present. The memories of industrial workers' refusal of labor toxicity intertwine with everyday experiences of recalcitrant social natures to sustain practices of commoning that challenge both private and public ownership.

Inaugurated in 1923, the industrial complex used highly toxic carbon disulfide for manufacturing artificial silk. A site of capitalist discipline, chemical exposure, and workers' insubordination to a rigid organization of labor, the factory operated until 1955. For decades, it remained abandoned, the industrial buildings left to time and the elements. Then, in the early 1990s, a property tycoon unveiled plans for building a shopping mall. Soon, the project moved to construction but the excavators struck an underground aquifer. Despite drainage attempts, the water kept rising, flooding the construction site and making it impossible to proceed with the development plan. Eventually it stabilized, forming the urban lake now bordering the factory ruins. In the years that followed, it altered the landscape, creating the conditions for the development of a highly biodiverse ecosystem embedded in a larger urban matrix (Battisti et al. 2017).

Soon after the lake's formation, local residents and activists started to reclaim the area. At first, they occupied some of the abandoned buildings and opened a social center, a laboratory for culture and politics (Mudu 2004). Then, a broader alliance of neighborhood associations negotiated the creation of a city park, the Parco delle Energie, managed by a forum bearing the same name beginning in 2008. A form of governance from below, the forum meets once a month as an open assembly that makes collective decisions about the park and the lake. A group of people established an archive, the Maria Baccante Documentation Center, collecting documents found in the abandoned factory and sustaining a process of collective memory about living and struggling in a chemical infrastructure. Contesting the collusions between real estate developers and public administrators, the forum has been reclaiming the factory ruins and the lake as commons, that is, a site of cooperation and a mode of living alternative to the pursuit of profit. In 2014, when new plans for residential construction were announced, the forum launched a mobilization to oppose the project. Struck by how the lake had transformed the landscape, the activists began referring to the body of water as the "lake that resists," evoking its power to act.

Drawing on the philosopher of science Isabelle Stengers, I argue that what has been emerging at Ex-SNIA is a *cosmopolitical commons*, an experimental project brought into being by the "becoming-with" of the activists, the lake, and the memories of vulnerability and resistance to the toxicity of factory work.[1] As I showed in Chapter 3, influential critical theorists frame the commons as the product of human cooperation that has the potential to generate radical projects of self-governance (Hardt and Negri 2009; Harvey 2012; Dardot and Laval 2014; De Angelis 2017). Even urban political ecology perspectives, attentive to the dynamic metabolic flows that make up the city, ultimately have advocated for a return to the "properly political" that foregrounds human antagonism but leaves little space for the appreciation of other-than-human interventions in the political field (Swyngedouw 2014). Complicating these frameworks, in this chapter I join other feminist scholars to explore the imbrication of the social and the ecological in practices of commoning (Gibson-Graham and Roelvink 2010; Tsing 2015; Singh 2017; Clement et. al. 2019). I contend that the Ex-SNIA provides a glimpse into a situated socio-ecological commoning that strives to persist in a Southern European urban context marked by the increasing precarity of labor and livelihood. Memories of struggle, stories of toxic embodiment, and bodily resistance to the toxicity of labor function as intergenerational reverberations arising from the postindustrial ruins of the Ex-SNIA.

In order to examine this multilayered struggle, I build on a robust body of work in the feminist environmental humanities, political ecology, and science studies. Feminist scholars as diverse as Vanessa Agard-Jones (2014), Stacy Alaimo (2010), Stefania Barca (2020), Mel Chen (2012), Michelle Murphy (2008), and Julie Sze (2018) have all engaged with questions of toxicity, drawing attention to the interactions and the uneven distribution of power linking a variety of asymmetric agencies, both human and nonhuman. By foregrounding the environmental injustices that affect bodies hierarchically arranged by structures of race, gender, class and disability, this scholarship contributes to "counter-hegemonic narratives" about socio-environmental conflicts (Barca 2014).

Working through these theoretical and methodological insights, this chapter weaves together archival research, interviews with activists, and sustained attention to the force of the nonhuman entities that compose the liquid landscape of the Ex-SNIA. I begin by sketching a brief history of the SNIA Viscosa as a chemical infrastructure where the fascist governance of populations and biological life coalesced with the political economy of chemicals to produce forms of toxic embodiment (Cielemcka and Åsberg 2019). Then, I discuss the Ex-SNIA's reconfiguration as commons with cosmopolitical qualities, an unfolding project of collective memory and invention enacted by activists through a process of becoming with nonhuman actors and forces. Through the in depth analysis of this political experience, I continue to explore how the contemporary resurgence of the commons transforms dominant regimes of property and governance while at the same time altering approaches to the commons that rely on the distinction between active human collectives and malleable resources.

THE EX-SNIA AS CHEMICAL INFRASTRUCTURE

Following the feminist scholar of techno-science Michelle Murphy, by "chemical infrastructure" I mean the spatial and temporal distribution of chemical agents as they circulate through systems of production, consumption, and waste, producing effects on bodies and landscapes. As volatile chemicals move within infrastructures, they permeate human and nonhuman beings, producing forms of "chemical embodiment" (Murphy 2008). Chemical infrastructures comprise physical systems and social sedimentations that bind together humans and nonhumans. Writing on toxicity, reproduction, and time in relation to Native populations in Canada, Murphy notes that social sedimentations include the legacies of settler colonialism, capitalist relations, and hierarchical distinctions of race and sex. I understand the Ex-SNIA as a chemical infrastructure built through the appropriation of nature and the exploitation of labor (Moore 2018). This enclosed and yet highly porous space produced forms of chemical embodiment connecting state structures and imaginaries, gender asymmetries, and transnational economies (Murphy 2010). In the factory, chemicals circulated within and outside closed walls, through bodies and machines, air and water, in a space where the capitalist organization of labor intersected with fascist governance. This case study suggests the importance of understanding chemical toxicity as deeply connected to the toxicity of labor within historically situated ways of governing the life of populations.

Murphy focuses on the afterlife of chemicals, their wildly uneven temporality, and intergenerational effects on particular populations. Some chemicals elicit immediate responses in human and nonhuman organisms. Others dissolve and still others produce a slow and delayed violence whose causes are difficult to trace (Nixon 2011). The attention to latency and time is invaluable for tracking the differential distribution of chemical violence across populations. But Murphy is also interested in a different kind of intergenerational reverberation, that is, the use of memories of the past for creating more just futures. She asks, "Which pasts need to be pulled out of the sediment into activity? What pasts can be drawn into new action?" (Murphy 2013). I take up this question to examine the role played by memories of toxicity and workers' struggles in ongoing activist practices at the Ex-SNIA. By doing so, I trace how this chemical infrastructure became a field of commoning.

On September 5, 1923, the siren screamed out for the first time, calling workers to enter the new rayon plant in Rome's Prenestino-Labicano neighborhood (Sotgia 2003). One of the few industrial areas in the city, the artificial silk factory was built on a liquid landscape. As the Ex-SNIA activists found out by studying hydrogeological maps from the late nineteenth century, a water stream, known as Fosso della Marranella, used to flow there, excavating tunnels in the volcanic rock, and then merging with the Aniene River. The plant appropriated the water by funneling it through a pipe, used it in the viscose manufacturing process, and then dumped chemical waste back into the stream and the land. This was the same aquifer body that in the early 1990s would erupt from underground to feed the lake and stand in the way of urban speculation.

The Prenestino-Labicano neighborhood, located outside the city walls but close to the railroad that transported workers and materials, was populated by a cheap workforce, people living in a chaotic aggregate of brick huts and shacks. This was one of the peripheral slums established as a result of Mussolini's policy of forced resettlement. Starting in 1924, thousands of houses in the city center were demolished, their inhabitants forcibly relocated far away from the centers of political power (Insolera 1962). Prenestino-Labicano was home to displaced people and migrants from Southern Italy, small artisans, peasants, the unemployed, the poor, and petty criminals. Pier Paolo Pasolini's *The Street Kids* (2016) describes it as crowded by "peasants from Puglia or the Marches, Sardinia or Calabria, young and old . . . returning drunk and ragged." Kids filled the streets "disorderly as a swarm of flies above a dirty table" (ibid., 91–92). For many people living there, the new rayon factory was the only source of wages.

Initially the factory employed about 2,500 workers, half of whom were women, a figure that went up to 70 percent between 1938 and 1955. Labor was organized along gender lines. The chemical department employed a mainly male workforce while women, many of them as young as thirteen years old and paid much less than the men, were the majority in the textile plant. On a hill next to the factory there were workers' dorms and other services, including a grocery store and a small daycare. The management organized a variety of courses and recreational activities. Women practiced sports for improving their health, attended sewing lessons and other courses designed to foster the development of "female capacities" (Nerbini 1925, 161). These policies were in line with fascist programs of social assistance aimed at neutralizing class antagonism by creating relations of dependency linking workers to the factory (Tommolillo 2020). For Mussolini, "Smart capitalists are not only concerned with salaries but also with housing, schools, hospitals and playgrounds for their workers" (Nerbini 1925, 45). The production of wealth had to be integrated within the biopolitical control of populations under the nationalist state. The establishment of social services within the factory served the purpose of better controlling the workers by tightly organizing their lives.

Chemical agents were ubiquitous in SNIA workers' everyday life. Carbon disulfide, an ingredient in economic processes and practices of state-making, created distinctive forms of toxic embodiment. First synthetized by a German chemist in 1796, this substance had been used in industrial manufacturing since the 1850s, initially in the cold vulcanization of rubber for making condoms and balloons, then as a widely promoted pesticide and a key ingredient in the viscose process that turns wood cellulose into rayon. Its effects on human bodies range from sensory damages to debilitating neurological symptoms. They include fatigue, irritability, hallucinations, depression, and suicidal impulses. Prolonged exposure also causes a high risk of heart disease and parkinsonism. As Paul David Blanc (2016) demonstrates, carbon disulfide poisoning in industrial settings became more frequent with the rapid expansion of the rayon industry. Already established before World War I, the viscose industry boomed in Europe, the United States, and Japan in the interwar period, becoming one of the first multinational corporate enterprises. Through international exchanges of technical knowledge and patent agreements between Europe and the United States, it gave rise to multinational industrial giants. Although industrial development was late and slow in Italy, the Italian rayon industry grew so fast that by the mid-1920s it was second only to the United States in terms of quantities of rayon produced, and it was first in terms of exports. SNIA Viscosa was the leading firm in the business, fully participating in its transnational expansion.

Founded in 1917, the Società di Navigazione Italo Americana (SNIA) was initially as a company transporting coal from the United States to Italy. After World War I, it entered the chemical-textile industry, quickly becoming Italy's powerhouse in the sector. As a firm exporting the large majority of its production, SNIA Viscosa was heavily hit by the revaluation of the lira in 1926 but it fully recovered in the following years. In the mid-1930s, it benefited from the fascist policies of economic self-sufficiency, adopted by the regime in reaction to the sanctions that the League of Nations placed on Italy in the aftermath of the colonial invasion of Ethiopia. Rayon played a significant role in the fascist imaginary of modernity, becoming identified as a product conjoining land, the renewed Italian spirit, and technological power. Commentators close to the regime heralded the SNIA factory in Rome, and the rayon industry as a whole, as a symbol of the dynamic Italian spirit, an icon of modernization, economic progress, and technological innovation (Nerbini 1925). According to Karen Pinkus (1995), artificial fibers, particularly rayon, were "a central obsession for the regime" (ibid., 213). Advertised as a fabric available to rich and poor, manmade rayon fit well with the interclassist, technocratic, and masculinist vision of the fascist nation. It replaced delicate and expensive silk, a material marked as feminine and exotic. Used for manufacturing uniforms and backpacks for Italian soldiers during the colonial war in Ethiopia and World War II, rayon came to be associated with militarized masculinity at a time of imperial ambitions. Together with steel and plastic, artificial textiles were identified with the form of embodiment of the new fascist man. As the literary critic Jeffrey Schnapp (1997) writes, these materials became "a site for elaborating a complex physics and metaphysics of sovereignty that celebrated, on the one hand, a limited and limiting national/natural landscape (imbued with attributes of heroism and moral superiority) and, on the other hand, the unlimited power of technology, culture, and the national will to transform that very lack into abundance, beauty, and strength" (ibid., 195). The symbolic investment into artificial textiles corresponded with the robust state support for the industry. Rayon, defined as "the most modern of Italian fabrics and the most Italian of modern fabrics" (Maxwell 2014, 127), was one of the newly developed materials fueling Mussolini's drive toward economic autarchy. In turn, autarchic policies bolstered the domestic consumption of artificial silk, thus benefiting the SNIA Viscosa.

The power of chemicals and machines inspired the futurist poet Filippo Tommaso Marinetti, an ardent supporter of the fascist project, to write several poems celebrating the transmutation of trees into artificial textiles. In the viscose industry, the poet, guided by "the haunting sickly-sweet scent of sulphur," witnessed a "molecular drama" (Marinetti, quoted in Schnapp 1997, 243) in which the furious mingling of raw materials, human labor, chemical baths, and weaving machines produced "distinct mechanical and chemical personalities" (Schnapp 1997, 209). Marinetti imagined rayon as a second skin, "the technologically enhanced double of a primary skin seen as once enfeebled and enervated but reinvigorated thanks to the fascistization of the Italian body politics" (ibid., 197). As chemically transfigured organic matter, rayon and other artificial fabrics were "the living prolongations of living beings" (ibid., 240), a compound of land, chemical reactions, and national spirit. In the futurist imaginary, chemical personalities were technologically enhanced "heroes to be praised and sung" (ibid., 209), figures of militarized masculinity capable of overcoming nature and transcending bodily limits. The chemical embodiments emerging out of viscose factories, however, diverged profoundly from fascist superhuman fantasies.

In the Prenestino-Labicano factory, production was organized through the Bedaux system, a method for speeding up work and enhancing managerial control. Carbon disulfide was released into the air, traveling in the poorly ventilated factory from one room to another, leaking into the porous bodies of workers through inhalation or skin contact.[2] The effects of toxic chemicals combined with the toxicity of factory work, the impact of long hours, repetitive tasks, machine noise, and strict managerial supervision and punishments. In the mid-1920s, international medical literature started reporting the effects of carbon disulfide on artificial silk workers. In Italy, Giovanni Loriga, chief medical inspector for the Ministry of Labor, published the 1925 investigation titled "Hygienic Conditions in the Artificial Silk Industry." The report described unwholesome working conditions and detailed the risks of carbon disulfide intoxication for each phase of the viscose process. Loriga mentioned "an epidemic of nervous disease" affecting women working in the fiber twisting department. Some physicians, he added, diagnosed the outbreak as "collective hysteria, while by others it was judged to be due to carbon disulfide" (Loriga 1925, 89).[3] Blanc (2016) notes that Italian occupational medicine was one of the most advanced in the study of carbon disulfide intoxication (ibid., 38). However, this needs to be understood within the larger context of occupational medicine under Mussolini. While physician-patient relationships became more limited and bureaucratized, the fascist support for research in the field was part of the effort to increase workers' efficiency in a factory regime seen as indispensable to the growth of the nation. This shift was eloquently evoked by a prominent occupational doctor at the Seventh National Conference of Occupational Medicine: "doctors have to trust more than ever a government that has full knowledge of the value of life and its productive power" (quoted in Carnevale and Baldasseroni 1999, 71).

Many viscose workers from the SNIA factory were brought to the Policlinico, a large university hospital, with symptoms of chemical poisoning including psychosis, mood changes, hallucinations, and fatigue. Working at the Policlinico, the prominent physician Aristide Ranelletti (1934) reported several cases of carbon disulfide intoxication. An occupational medicine expert, Ranelletti (1934) described his work as aiming "to induce industrialists to improve labor systems" (ibid., 73). In the fascist state, occupational medicine was embedded in a biopolitical project aimed at reducing risk both for building consensus and increasing national productivity within the scientific organization of labor. According to the historian Francesco Cassata (2011), this field was informed by a constitutional approach that placed emphasis on individual psychophysical predispositions to illnesses rather than toxic exposure in the workplace. Accordingly, each worker had to be placed in the profession that minimized the risk of contact with toxic and infective agents to which he or she was already sensitive. Constitutional medicine underpinned the introduction, in the late 1920s, of preventive measures for workers exposed to hazardous agents. People with nervous system issues, for example, were exempted from working in the viscose industry. Additional measures were also introduced that insured workers of chemical departments diagnosed with carbon disulfide intoxication. Such legislation was extended to include workers from textile departments only in the mid-1950s (Sotgia 2003, 199). By then the rayon factory in Prenestino-Labicano had already been shut down.

In her study of carbon disulfide exposure at the SNIA factory in Rome, historian Alice Sotgia asks whether workers knew about the hazards. Crossing records from the factory and the psychiatric hospital Santa Maria della Pietà, she identifies several cases of poisoned viscose workers stretching from 1927 to 1940. Sotgia reports the case of a thirty-five-year-old man who was hospitalized in 1927. Although the doctors did not identify carbon disulfide as the cause of the patient's acute paranoia, he associated his symptoms with poisonous gases inhaled in the factory. Another man, who had been hospitalized in 1929, "complained he had been poisoned by acids" (Sotgia 2003, 205). This suggests that workers were aware of health risks from chemical exposure in the factory.

These risks, however, were part of an experience of work where toxicity extended beyond the effects of chemicals, to the capitalist organization of labor under the fascist regime. Workers, water, trees, and other raw materials were vital resources in the viscose productive process. The chemical violence in the rayon industry reduced workers to a kind of waste that was discarded and moved from one enclosed space to another, from the factory to the hospital (Foucault 1995).

Chemical violence on SNIA workers has been widely documented. Ample evidence is available at the Ex-SNIA archive.[4] Yet no clear indication remains of industrial contamination of water and soil linked to the chemical-textile complex. It is possible that carbon disulfide and other chemicals dissipated in the stream of water and dissolved in the soil and the air. The urban lake that emerged from underground decades after the factory's closure bears no trace of past toxicities. In this sense, the latent, delayed violence that defines chemical infrastructures, shaping the future of the people and the other-than-human beings inhabiting them, does not seem to be central here. What has been emerging at the Ex-SNIA, I suggest, is a different kind of intergenerational effect, one made possible by a political project that weaves together resistant memories of labor toxicity with the narrative of the rebel lake.

THE ARCHIVE OF ECO-MEMORIES

At the SNIA Viscosa, the effects of industrial chemicals were not limited to the enclosed space of the factory. Carbon disulfide and other chemicals traveled outside the factory walls, permeating urban spaces and bodies. Thus, they produced various instances of chemical embodiment. In one of several conversations during my visits in Rome while researching this book, Matilde Fracassi, a retired educator and activist with the Forum Parco delle Energie, provided an account of transient olfactory sensations related to the factory. She recalled, "I have always lived in this neighborhood. As a child I would sense the smell of the factory brought by the North wind. It wasn't particularly strong or unpleasant, it was a sugary smell, probably caused by chemicals. It was a presence."[5] Carried by air, chemical agents were volatile entities that infused the urban landscape and deposited in bodily memory. Although Fracassi never set foot in the factory, smelling chemicals was part of her childhood experience of the city. She continued by mentioning other forms of chemical embodiment: "A local hairdresser told me that she would guess which clients worked at the factory by smelling their hair. She could sense the chemicals." Chemical sensations, whether experienced directly or through other people's stories, assume a new significance in the context of Fracassi's participation to ongoing struggles. As Murphy (2013) points out, "Quotidian acts of breathing, drinking, and smelling can become knowledge-making moments in chemical infrastructures." Fracassi's sensory interaction with industrial chemicals contributes to a collective memory that brings together human and nonhuman entities. Such olfactory engagement participates in the making of an archive of eco-memories that has emerged out of the struggles within an expanded chemical infrastructure. The memories of smell enrich the archive, extending the bodily experiences of the past into the present and creating intergenerational links between the political struggles of factory workers and current activism.[6]

After the factory shut down in 1955, Matilde could see the walls and the top of the trees from her apartment window: "It was like seeing the woods, dark and thick with trees, a world of magic and mystery." She entered this world only many years later, in the early 1990s, when a real estate developer named Antonio Pulcini unveiled the plan to convert it into a shopping mall and activists reclaimed it as a collective space. At the time, the Prenestino-Labicano was a somewhat run-down neighborhood where formal, informal, and illegal economies overlapped. Former SNIA workers mingled with people who had been involved in 1970s radical movements, younger students facing labor precarity, the unemployed, and a variety of people skilled in the art of making do. As Alessandra Conte, a historian and activist told me, some local residents had begun to explore the industrial ruins and were scavenging for what they could: glass, metal tubes, and other materials to reuse. In an office building, they found documents left behind by the factory management that became the core of the Ex-SNIA archive.[7]

Officially opened in 2012 in a former SNIA building renovated with public funds and now part of the Parco delle Energie, the archive has been the labor of love of a group of historians, archivists, urbanists, and local residents, many of whom are also members of the Forum Parco delle Energie, the assembly of activists reclaiming the Ex-SNIA. The archive collective has been organizing, studying, and making accessible piles of dusty documents from the rayon plant including maps and pictures, administrative papers, medical records, and workers' files. My research at the archive as well as conversations with the activists shows that the medical records report cases of suspected carbon disulfide intoxication in male workers between 1930 and 1938.[8] This confirms that although the factory administration was aware of the problem, it was reluctant to admit its extent and take action to prevent consequences for the workforce. The medical notes, handwritten by factory physicians, report workers' descriptions of fatigue, impotency, hallucinations, and depression. Some workers collapsed on the job, and others were irascible and prone to anger with supervisors. Several were hospitalized. Many were sent to the INFAIL, the industrial accident insurance fund instituted by the fascist regime, which often dismissed the suspected diagnosis, thus rejecting insurance claims. For example, a worker, whom I will identify as A. B., reported fatigue, impotency, and mental confusion in January 1938. The INFAIL sent him back to the factory noting that he presented "no signs of professional illness."[9] Three months later, he showed up again at the factory medical office with his wife. According to medical records, she stated that he "would tell her rambling stories that scared her." A. B. was transferred to the hospital and from there to a psychiatric ward for treatment for depression and hallucinations. He went back to work in September of the same year.

At the archive the records of individual workers are preserved in old orange folders and binders, some still smelling of dust and mold. For each worker, the management collected information about level of instruction, work experience, family information, general health conditions, and behavior on the factory floor. The managerial control over the workforce combined with fascist surveillance over individual morality and political orientation. After the regime introduced racial laws in 1938, the factory was required to keep a certificate of membership to the "Arian race" and the identification as Catholic for each worker. This was part of the effort to monitor and govern conduct through a capillary disciplinary apparatus. But the archive also contains traces of counter-conducts and ways to disrupt the flow of production. In the orange folders, one finds detailed lists of punishments. They reveal that workers were fined, suspended, and fired because they were too slow, too sloppy, or undisciplined. Some were punished for insubordination, others for chatting too much and chanting aloud, for being late, washing hands, smoking, and taking breaks without permission.

Produced as a means of monitoring workers, this body of papers acquires different meanings in the hands of activists who are keenly aware of the power-laden relations involved in the constitution, interpretation of, and access to the archive (Derrida 1996, 4). By connecting managerial files with larger urban histories, the archive functions as an incubator for memories of struggle buried in the unconscious of the city. The rayon factory and the Prenestino-Labicano neighborhood had been the scene of labor disruption and anti-fascist resistance. The first major strike occurred in 1924, launched by eight hundred women workers demanding salary raises and the improvement of working conditions. In 1949, the workers occupied the factory for over a month as a protest against layoffs. Once again, women were at the forefront of this action, which was widely supported by neighborhood residents.[10]

Not by chance, the archive is named after Maria Baccante, a former partisan and factory worker who had a leading role in the 1949 protest. Conte, who spent years reconstructing Maria's life, explains, "She kept coming up in our research as if she was demanding attention. It was impossible to ignore her."[11] Originally from L'Aquila, raised in a well-off household, Maria moved to Rome on her own, perhaps to escape suffocating family rules. She never got married; instead, she became a partisan and participated in the armed struggle against fascism, thus contradicting the entrenched notion that women had mainly secondary roles in the war of liberation. In 1947, Maria started working at the SNIA, one of few women employed in the chemical department. As Conte points out, "Her archival file shows that for four months in a row she asked for new shoes, probably because they were corroded by chemicals. Her file is full of insubordinations and punishments. She was not alone in this, there were many Maria Baccantes in the factory."[12]

At times, workers' contestation of managerial control acquired a striking bodily quality. In one case, a young woman whom I will identify as M. M. was caught urinating on the shop floor. A report in her file reads, "not only did she admit the fact without shame and regret but she also showed animosity toward the bosses who deny workers everything, even permission to use the toilet. From a complex of elements, it appears that the action was undertaken as a manifestation of a general intolerance of discipline."[13] The young woman turned improper bodily exposure into a means of insubordination. She enacted bodily resistance against the disciplinary regime of the chemical infrastructure. The materials available at the archive do not point to workers' protests specifically addressing chemical poisoning, but they provide insights into embodied forms of resistance to the toxicity of labor in this chemical infrastructure. The embodiment of toxicity produced not just victims but bodily instances of refusal and disruption of the smooth flow of production.

Maria Baccante and M. M. are ghostly presences that exert a powerful force on the activists. These haunting figures evoke instances of insubordination and refusal whose power has been revitalized in the present. As Matilde told me, "As I researched the workers' files, they began talking to me. Their voices help us to protect this area from speculative interests. There is a common thread running through their struggles and ours."[14] But what is the common thread? Does it make sense to speak of commonalities given that, at least in Western contexts, the Fordist factory, the paradigmatic model for the organization of labor in the industrial age, has been largely, although not always, relocated to other places and supplanted by forms of precarious labor that blur the divide between work and life? At the Ex-SNIA histories of toxic embodiment and bodily resistance to the toxicity of labor contribute to intergenerational memories of struggle. Through the effort to sustain collective memory, the local archive connects local critiques and knowledges from below developed within the chemical infrastructure of the factory and the neighborhood (Foucault 2003). In the archive, activist research, enriched by countless conversations with former rayon workers and elderly neighborhood residents, intersects with scholarly work on labor history and the effects of toxic exposure. Taken together, these perspectives allow them to "constitute a historical knowledge of struggles and to make use of that knowledge in contemporary tactics" (ibid., 8). The intimacy with past contestations of labor toxicities made possible by the Ex-SNIA archive bespeaks the possibility of refusing the capitalist organization of labor and life at a time when the doctrine that "there is no alternative" to capitalism has taken hold of the contemporary imagination. Such intimacy, moreover, may be useful to raise questions about new forms of toxic labor and chemical infrastructures. If in Western postindustrial cities the factory is no longer the heart of the productive process, what other forms of chemical embodiment and toxic labor have emerged in the European social factories?[15]

The political potential of the Ex-SNIA project of collective memory is enhanced by the ways in which it brings together the past of the factory and the present of the lake, industrial history and the natural sciences. The activist work at building and maintaining the archive is intertwined with that of ecologists, engineers, and botanists who have been producing and sharing knowledge from below about the lake's water quality, the geology of the area, and its rich biodiversity. Rosanne Kennedy (2014) proposes the concept of "eco-memory" to refer to collective forms of remembrance that, in the context of Indigenous histories of dispossession in Australia, link together human and other-than-human beings in narratives of vulnerability and survival. To be sure, the memories and narratives that are elaborated by the Ex-SNIA should not be conflated with Indigenous eco-memories of interspecies kinship. Kennedy's concept, however, is valuable for thinking about how labor history, chemical interactions, and the transformation of urban ecologies are layered into the memory of the postindustrial site comprising the factory and the lake. As Flavia Sicuriello, an ecologist and activist, argues, "Creating narratives that intertwine the histories of viscose industry workers with the life of plants and birds is what makes this mobilization so powerful."[16] Crafted and shared through a variety of activities including guided tours, urban beekeeping, projects with schools, bird watching, picnics, and conferences, these narratives convey the "experimental togetherness" (Stengers 2005) of the Ex-SNIA as a dense urban ecosystem, a site of contestation and affirmation. In this context, eco-memories are sources for political action, intertwining with forms of conviviality and self-organization. The lake, a recalcitrant entity capable of standing in the way of urban development, has been a central figure in this project.

THE REBEL LAKE

At first the lake looked like a post-disaster zone. Pictures from the early 1990s show a massive excavation partially filled with water, clogged with debris and construction materials. The frame of a building, with scaffoldings wrapped in mesh, stuck out into the water. Conte told me, "It was ugly, it showed the violence on the environment and the reaction to that violence, the eruption of water that forced the closure of the construction site."[17] The flooding is seen as the environmental response to the untenable pressure of real estate development. The landscape was devastated. Then, it slowly changed. The water level rose and stabilized; plants began to grow inside and around the lake; fish, birds, and foxes showed up and stayed. The fenced property became a land of adventures for the neighborhood kids. Ecological repair was a slow-moving,[18] steady process leading to the formation of a highly biodiverse habitat that, according to some researchers, includes about three hundred plant and sixty-two bird species (Battisti et al. 2017). Quietly, the lake, whose waters continue to move and flow, began to exert its force of attraction on the many who learned to care for its existence.

In 2013, the activists assembled in the Forum Parco delle Energie learned that the landowners had a new plan for the construction of a residential complex in the area. When studying the development project, which had been leaked to them by a municipal politician, the forum found out that part of the land had been expropriated in 2004 for public use. This meant that no construction was allowed. But there was a problem. According to Italian law, if expropriated land is not utilized within ten years, it must be returned to the previous owner. After months of protests and grassroots events, journeys through a maze of bureaucracy, and negotiations with the city, the activists came close to a victory. The city administration, then guided by a center-left coalition, pledged to block the development plan and work with the forum. It even allocated money for equipping the area around the lake and making it accessible to urban dwellers. The funds, however, were quickly diverted elsewhere. Since April 25, 2016, the lake area has been accessible thanks to the collective organization of the activists who keep laying claims to the place.[19] After a popular consultation among visitors and activists, the body of water has been named "Bullicante Lake," a reference to the Roman name for the effervescent waters once present underground. Visitors are welcomed by signs explaining the basic principles and rules governing the place. It reads, "This is a place liberated from profit and from real estate speculation, thanks to the participation and the struggle of everyone. It is a place that lives from self-management, self-financing, and solidarity." The sign lists prohibitions, like swimming in the lake and leaving litter on the ground, as well as some encouraged activities, including "playing ball, getting dirty, laughing out loud and shouting with joy." While in 2020 a large portion of land around the lake had been officially declared a "natural monument," and enjoyed protected status, as of this writing in August 2025, about one third of the Ex-SNIA complex remains private and is thus at risk of being covered in concrete.[20] The struggle is ongoing, and it has been unfolding on the terrain of the commons while the relationship with public administrators has been fraught with frustration about hollow promises.

In Italy, the resurgence of the commons can be roughly traced back to 2011, when a large grassroots coalition used the commons as the key building block in a successful voter referendum to repeal water privatization. From water, claims for the commons have extended to urban and rural spaces, cultural practices, and technologies. A diverse network of activist groups has been exploring the many dimensions of the commons: its history and legal aspects, its role in contemporary advanced capitalism, and political potential for unsettling state sovereignty and private property while at the same time creating alternative modes of governance from below. In Rome, from 2011 to 2014, the Teatro Valle, an eighteenth-century theater at risk of privatization was occupied by artists and technicians and became internationally renewed for merging performing arts with the project of designing new legal architectures for the commons (Pinto et al. 2014; Capone 2020). This experience inspired many other collaborations between workers, artists, scholars, and activists advancing theories and practices of commoning. In Milan, from 2012 to 2021, the autonomous cultural center MACAO experimented with alternative economies and technologies of care, including the cryptocurrency Common Coin, designed to support cooperative projects (Terranova and Fumagalli 2015). In Naples, the coalition Rete Commons opened urban parks and reclaimed land once owned by criminal organizations for creating an agricultural cooperative while also animating struggles against toxic contamination (Armiero and De Angelis 2017; Armiero and Sgueglia 2019). Through assemblies, publications, conferences, performances, and a range of other initiatives, activist groups have reactivated the commons as a conceptual framework and collective practice resisting neoliberal governance based on the pursuit of individual realization in a context of increasing labor precarity and the dismantling of public services. It is against this rich backdrop that the Ex-SNIA's struggle has been unfolding. Issues of property and conflictual interactions with both private actors and public institutions have been at the heart of this project.

As Matilde Fracassi has aptly noted, "the lake has confounded the boundary between public and private property."[21] In Italy, deep waters are inalienable;

they cannot be bought or sold as they belong to the state. The Forum Parco delle Energie has been putting pressure on the city council and the regional government, demanding the recognition of the lake as public domain (*demanio*) and the institution of a protected area including the factory and the watery formation. On the one hand, the activists are proposing to establish the primacy of the public over private real estate interests. They use the public as a platform for building alternatives to the logic of the market. On the other hand, they have been experimenting with enduring and evolving practices of assembling, reclaiming, and collective decision-making that point to the commons as an alternative to both state power and neoliberal regimes of governance. In doing so, the Ex-SNIA activists have been pushing for the radical reorganization of public institutions through claims for participatory decision-making.

David Harvey observes that radical groups often resort to a mix of property arrangements and modes of governance in order to procure access to space and keep it off limits to commodification and market exchange. Insofar as the public is often the expression of state power working in alliance with market forces, activists' relationships with it can be highly complicated. The reclaiming of public goods and resources, however, can provide an opening for the creation of new modes of commoning arising from political action. As Harvey (2012) observes, the making of the commons "are contradictory and therefore always contested" (ibid., 71). In Harvey's nuanced account, commoning is presented as a process situated in particular places and constrained by neoliberal forces that at times use decentralization as a vehicle for deregulating markets and increasing inequalities. Harvey's work is also notable because it openly argues against the divide between "natural commons" (water, land, and forests) and "social commons" (ideas, images, code) on the ground that "all resources are technological, economic, and cultural appraisals, and therefore socially defined" (ibid., 72). This position resonates, at least in part, with the Marxist approaches that I have explored in Chapter 3, including Michael Hardt and Antonio Negri's (2009) definition of commoning as the cooperative social interactions that constitute both the field of capitalist capture and the terrain for building alternatives to it. But Harvey's work on urban commons is distinctive in that it builds upon his long-standing engagement with the notion of metabolic relation to nature and space.

Rejecting any separation between nature and society, Harvey (2000) contends that humans "are sensory beings in a metabolic relation to the world around us. We modify the world and in doing so change ourselves through our activities and labor" (ibid., 207). Drawing on Marx's use of the term *Stoffwechsel*,[22] Harvey understands metabolic relations as the flows of energy and labor that link and transform both humans and nature. As a social formation, Harvey observes, capitalism rests on a metabolic relation with nature; it shapes the environmental conditions of its own reproduction, but it does so in a context where autonomous evolutionary forces are also constantly reshaping environmental conditions. Several Marxist scholars of urban political ecology have engaged these issues, rejecting the binary opposition between urban and natural environments and proposing the image of the city as hybrid of human and nonhuman, social and biophysical processes, human labor and the earth (Heynen et al. 2006). This field has produced empirically grounded analysis of "how social forms of power transform the environment and how the nonhuman becomes actively enrolled in processes of uneven and combined socio-ecological production and re-production" (Ernstson and Swyngedouw 2019, 12). While I value these insights, I remain unconvinced with analysis of metabolic relations that look at nonhumans as being merely "enrolled" in processes of socio-ecological reproduction. Still more, I am wary of how scholars of urban political ecology call for unsettling the postpolitical techno-managerial governance of cities and natures through a return to the "properly political" (Swyngedouw 2014). Even as I concur with the criticism of a postpolitical condition that replaces conflicts around socio-ecological configurations with managerial planning and technological fixes, I worry that focusing on a return to the "properly" political forecloses the possibility of opening up established understandings of what counts as political. Specifically, I am concerned with the reconfigurations of the political and politics that account for the ways in which other-than-human beings force thought and actions.

To clarify my position, I draw on Isabelle Stengers's (2005) articulation of *cosmopolitics* as a concept that troubles Western understandings of politics as a gathering of bounded human subjects capable of making judgments in the interest of the community. The prefix "cosmos" is meant to make explicit the presence of disparate entities and beings that do not have a political voice and yet come to matter politically. As I have shown in this chapter, the Ex-SNIA archive reactivates the memories of those who resisted labor toxicities in the factory. Their ghostly presences, I argue, combines with the force of the lake, which, rather than a resource around which human collectives coalesce, has been a prime actor in the activist mobilization. To be sure, the people "who build and sustain urban life" (Harvey 2012, xvi) have played a key role in the struggle directed at preserving the Ex-SNIA, the factory and the lake, as an accessible urban ecosystem. But human collectives are never acting alone. Ex-SNIA activists often describe the lake as an entity with "a soul," capable of effectively blocking real estate development. As Conte told me, the body of water "created a habitat that allowed us to fall in love with it." When the activists began regular visits to the lake, they "felt as if they were guests in a rarefied, unfamiliar, and regenerated space. These sensations translated into the imaginary of the lake that fights back."[23]

What does it mean to sense the lake as a regenerated space? And how do sensations translate into politics? In stark contrast with the image of the violence of urban development evoked at the beginning of this section, regeneration refers to the lake's capacity to transform urban landscapes in ways that have created propitious conditions for struggle. Embodied sensations of a more-than-human world can contribute to forming enduring *attachments*, what Stengers defines as "what cause people . . . to feel and think, to be able or become able" (Stengers 2005, 191). Attachments generate problems and pose questions, and they may propel transformations that could not be enacted by humans alone.[24] The lake, I argue, enabled attachments and collective actions that counter dominant regimes of property and governance.

The music video, *Il lago che combatte* (*The lake that fights back*), shot on location in Rome, visually represents the cosmopolitical experiment taking place at the Ex-SNIA.[25] Filmed in 2014, the video is a collaboration between the hip-hop band Assalti Frontali and the group Il Muro del Canto that combines rap as a form of urban storytelling with the tradition of protest music. The video crosscuts between two main locations, the concrete skeleton in the midst of the Bullicante Lake, and the rooftop of a building in the Prenestino-Labicano neighborhood. The video also incorporates footage of the SNIA ruins and street protests with the intergenerational coalition of residents struggling *with* the lake. The color palette is muted, tinted with the grey of concrete and the deep green of the water, punctuated with pops of red in the shots of protest actions. The song lyrics weave together past and present, recalling the everyday life in the factory, the murky practices of real estate developers, and the flourishing of more-than-human attachments. The encounter between a diverse group of people and the body of water that has repaired the postindustrial site is at the heart of this audiovisual production. As Militant A sings, "The lake invaded the reinforced concrete and asked for help / we learned to imagine, love, and experience it." This suggests a growing intimacy with the nonhuman environment. Later in the song, the lake is defined as "our lungs," thus connoting the transcorporeal dimension of a relationship connecting people and place (Alaimo 2010).

My intention here is not to suggest that the video conveys the image of an innocent nature that turns against plunderers to form commons that activists are called to protect. This would be a romanticized understanding of nature and activism. But the definition of the lake as "second nature" produced by capitalist development is not satisfying either (Smith 1984). While this concept captures the capitalist fabrication of nature for profit purposes, it runs the risk of eliding the ways in which second nature alters the beings that press upon it. It seems to me that Anna Tsing's notion of "third nature," that which "manages to live despite capitalism" (2015, viii), is more effective for thinking about the lake and the political project of which it is part. I understand the rebel lake as an instance of produced nature that manages to live in capitalist ruins. One could argue that hybrid spaces of experimentation are, to a certain extent, tolerated and even welcomed in the neoliberal management of urban spaces. Neoliberal capitalism, after all, has turned hybridity and entanglements between humans and nonhumans into central features of its operations (Pellizzoni 2025). But I propose that projects such as Ex-SNIA reject capitalist notions of nature, including urban natures, as that which can be made and unmade without consequences. In reclaiming the existence of spaces exempt from value creation and commodification amid uneven conditions of environmental and labor precarity, this project strives to provoke transformation in the form of the commons.

What I find striking is that in this process it is hard to tell where the natural ends and the social begins. While it does not make sense to think of the lake's water as natural commons in that its existence is fully integrated into the web of human activities (Mezzadra and Neilson 2013) it is likewise important to pay attention to the ways in which water flows, obstructs movements, dissolves some chemicals and reacts with others, enables the life of myriad beings and transforms landscapes (Braun 2004). The distinction between natural commons and social commons, I argue, breaks down not just because human activities and labor constantly mobilize material resources. Rather, what makes up the commons are dynamics in which other-than-human entities act as forces of socio-ecological transformation. By bringing into the fore some of these dynamics, the struggle unfolding at the Ex-SNIA points toward a more capacious sense of the commons than the dissolution of the natural into the social commons engendered by human interactions. Further, it invites us to reconfigure understandings of the political as a field of disagreement that cannot be enacted by humans alone.

This chapter shows that memories of toxicity and other-than-human actors have converged in the project of turning the Bullicante Lake into a site of commoning that counters tendencies toward urban privatization. The ghosts of past chemical embodiments and the narratives of worker's resistance to the toxicities of labor compose a frame of active memory that, combined with the force of the rebel lake, sustain what I call the cosmopolitical commons. By focusing on the cosmopolitical dimension of commoning, and thus foregrounding, and politicizing, attachments across time, species, and elements, this chapter complicates the prevalent understanding of the commons as the product of human activity. Thus, it further develops this book's overall argument that commoning is simultaneously social and natural, requiring the interplay of human and nonhuman actors for creating enduring, albeit still precarious, paths of decommodification.

6

CROSSING THE STORM: ENCOUNTERING DECOLONIAL PERSPECTIVES

In early May 2021, a group of Indigenous Zapatistas from Chiapas, Mexico, embarked in a journey across the Atlantic. The Escuadrón 421, made up of four women, two men, and a nonbinary person, set sail from Isla Mujeres, Mexico's most eastern point, in an old vessel named *La Montaña*. Hailing from the autonomous Indigenous territories formed in the years following the uprising of January 1994, the Zapatistas traveled to Europe to stage a "consensual invasion," a radical inversion of the colonial process that started with the conquest of the Aztec capital of Tenochtitlan, looted by the Spanish army led by Hernán Cortés. At the departure ceremony, the Zapatistas told the media that the journey would follow the route of the conquistadors in reverse, for sowing rather than destroying life. After fifty-two days, *La Montaña* docked in Vigo, Spain. This performance of "conquest in reverse" demonstrated that in the face of ongoing colonial violence and territorial dispossession, Indigenous people continue to exist and engage in the project of remaking the world. In a powerful challenge to the patriarchy and colonial gender politics, the first to set foot on European soil was Marijose, an *otroa*, a term part of the Zapatista lexicon that can be roughly translated as gender fluid (S. Marcos 2021).[1] Greeting the many activists gathered to welcome them, Marijose renamed the European continent "Slumil K'ajxemk'op" which means "rebel land" in Tzotzil, a Mayan language. The Zapatistas had come to Europe to tell collective stories of oppression and insurgence in Mexico, a country where militarized policing and drug trafficking generate widespread violence and insecurity for the population, particularly women, queer, and trans folk, the poor and Indigenous (Valencia 2018). They had come to share their transformative experience of social and ecological regeneration, to listen to people living, resisting, and making worlds on the other side of the ocean. Many more Zapatistas followed, forming a large delegation also including spokespersons from the

Indigenous National Congress. They traveled across Europe to meet with grassroots groups differently engaged in "the struggle for life," from defending territories against megaprojects to migrant support initiatives to climate justice organizations to feminist and queer movements and factory workers. The trip was supported by long-lasting networks of solidarity that in previous years had brought thousands of Europeans to Chiapas but rarely the Zapatistas to Europe.

The members of the delegation arrived in Rome in late October, holding meetings in activist spaces in the city. At a public event at the Bullicante Lake, the urban water body turned into a commons that I discuss at length in Chapter 5, a notable presence was María de Jesús Patricio Martínez (Marichuy), a well-known Indigenous *nahua* healer and the spokesperson for the Indigenous National Congress who had run in the Mexican presidential election in 2018. A key question on the table was the territorial mobilization against megaprojects supported by the Mexican government in the name of an economic development supposedly benefiting the population through the intensification of tourism, farming, and green energy production. In the same days, I participated in a nonpublic exchange focusing on women's and gender struggles where a group of Zapatista women met about one hundred feminist activists from Rome. Speaking in Spanish and Indigenous languages, they recalled the devastating effects of colonial racialization and dehumanization on their ancestors who had worked on the plantations. They also described how their political practices within the Zapatistas communities has addressed the specific oppression experienced by Indigenous women and *otroas* in colonial and postcolonial Mexico. Throughout an unusually chilly day, they spoke at length, responded to questions, listened to the contributions of a variety of collectives, and participated in a communal lunch in a disheveled urban space that had no heating but was brimming with energy, the colors of graffiti art, and photographs taken in Zapatistas communities.[2]

Each of these encounters was an elaborate choreography of translation, storytelling, and solidarity building. In preparation for the feminist gathering, participants were invited to read a short guideline about how to welcome and support the Zapatistas in ways conducive to mutual learning. An embodied activity opened the event, setting the stage for a generous exchange. The protocol of this and other encounters had been carefully crafted through weeks of preparation in which the Zapatistas posed specific requests about the length, the rhythm, the context, and the practices of conversation, including for instance avoiding drinking alcohol and taking pictures to circulate in social media. Even as they enabled mutual learning, they decided what kind of knowledge to share and when, at times making it clear that the questions posed by European activists, the concepts they used, were not entirely relevant for making sense of life and politics in the Zapatista autonomous territories. These moments of mutual hesitation, the misunderstandings, the narrative gaps, the efforts and failures to translate, were important as they made palpable persistent differences among the actors involved in assessing the effects of heterogeneous forms of dispossession and weaving alliances across disparate places.

I linger on these moments because they express the possibilities and difficulties of communication between "different onto-epistemic worlds" (de la Cadena 2015, 150), that is, between distinct modes of world-making in the context of neoliberal regimes of accumulation. In Rome, as elsewhere during their European tour, the Zapatistas recalled their commitment toward a politics of connected multiplicity as a response to the injustices of neoliberal capitalism, what they define as *un mundo donde quepan muchos mundos* (a world where many worlds fit). This vision of the pluriverse (Escobar 2020), and the engagement with Indigenous movements in the Americas more generally, have inspired much decolonial scholarship reflecting on questions of ontology, politics, and nature in ways that, more or less directly, interrogate the commons as a concept with European roots and global aspirations (Blaser and de la Cadena 2018). Although the encounters between the Zapatistas and the Italian activists in Rome did not focus specifically on the commons, practices of land use and forms of communal government were central to the conversations. Moreover, the "consensual invasion" was a unique opportunity of exchange between divergent ways of making worlds, thus providing an apt entry point for my discussion of the relationships between the commons and Indigenous communal practices. This chapter moves back and forth from Europe to the Americas, tracing how Indigenous movements complicate the understandings of what counts as politics that largely animate contemporary approaches to the commons. My analysis does not seek to approximate ethnographic detail but I am interested in how particular Indigenous struggles in Mexico and Bolivia have been interpreted by political theories of the commons and how these interpretations might be revisited to foreground the presence of nonhuman actors in insurgent collectives.

As a rich body of work in Indigenous and decolonial studies indicates, the Western understanding of what counts as politics is not a universal that applies across time and space. Rather, it has been part of the colonial project that claimed universality for itself. In the precolonial Andes, for instance, the Inca Empire was built through state practices that treated mountains, rocks, and the landscape as sentient actors involved in political processes (Wilkinson 2013). This was not limited to a vanished precolonial past but has survived through generations. Contemporary Indigenous groups in the Americas and beyond engage living and nonliving beings not as mere resources but as part of arrangements of existence (Povinelli 2016) or "political ontologies" (Blaser 2013; de la Cadena 2015a) that persist in spite of continuous dispossession. In the Americas, political ontological conflicts have come into sharp relief in the contemporary moment of intensified extractivism, when mining, drilling, and monocultural farming remove living and nonliving matter from the earth's depths and surface at an untenable speed and scale. Resource extractivism targets both territories and populations, and it violently reorganizes socionatural configurations of life and nonlife. In continuity with colonial capitalism that, starting in the sixteenth century, converted gold, silver, and rubber into global commodities, contemporary extractivism removes raw materials from the land and moves them through a range of infrastructural projects (Acosta 2010; Svampa 2019; Riofrancos 2020). Simultaneously, its reach extends to social ties and relations, deepening intersecting inequalities of race, gender, and class (Cielo and Cabo 2018). Extractive operations connect situated territories to global economies of labor and finance, extracting value from increasingly indebted populations (Gago and Mezzadra 2017; Mezzadra and Neilson 2019; Arboleda 2020). A central feature of this mode of accumulation has been the necropolitical indifference to the reproduction of territories and the communities of humans and nonhumans depending on them. This has become particularly clear in the case of Indigenous land treated as expendable. Native peoples, however, tend to "experience themselves in the world as having responsibilities to human, nonhumans and the environment" (Whyte 2016, 2). The recent flourishing of Indigenous struggles in the Americas—from the fight to keep oil underground in the Ecuadorian Amazon to conflicts in Bolivia over the construction of highways cutting through Indigenous land to the mobilization in Standing Rock opposing the construction of the Dakota Access Pipeline in North Dakota—points to what the Mohawk scholar Audra Simpson has defined as the Indigenous refusal "to go away, to cease to be" (Simpson 2014, 11).

While scholars of the commons have engaged with Indigenous movements, they often understand them as efforts to preserve and claim access to natural resources thus ignoring their political ontological dimension. This reflects what, in Chapter 3 of this book, I define as a paradoxical commitment of much commons scholarship to one of the key tenets of modernity: The distinction between humans as political agents and a natural world deprived of political relevance. In order to continue displacing the modern proclivities of current thinking about the commons, this chapter begins by attending to Indigenous critiques of universalist assumptions in prevalent approaches to the commons. Then, I turn to Indigenous struggles in Chiapas, Mexico, and Bolivia that have often been read through the lenses of the commons. Specifically, I examine Zapatista writings and political speeches that have brought the land and their ancestors into the fold of politics, and protest movements that in the early 2000s opposed the privatization of water in Bolivia on the grounds that water is more than just a resource, it is part of the fabric of life also comprising human beings. My analysis has at least two implications for thinking the commons. First, it invites caution in categorizing Indigenous struggles as struggles for the commons as this risks obfuscating the specificity of political ontological conflicts that only partially overlap with Western forms of commoning. Second, it draws attention to the presence of nonhuman actors in insurgent collectives that is often overlooked in commons scholarship, thus providing insights for reimagining the commons in light of modes of doing politics otherwise.

This chapter shows that Indigenous communal ways of living and struggling trouble conceptions of the commons that rely on the understanding of politics as a human affair. As a White European feminist scholar-activist, I enter into conversation with decolonial scholarship and activism with caution and respect, aware of the universalizing tendencies within Euro-American commons scholarship, and yet invested in the transformative potential of commoning. In addressing the encounter between the commons and decolonial communal living and thinking, I do not wish to claim the decolonial approach for myself.[3] Rather, I draw on generative decolonial perspectives for interrogating the blind spots of prevalent theories of the commons and exploring spaces of encounter between different, at times divergent, ways of relating to common use.

DECOLONIAL PERSPECTIVES AND THE COMMONS

As scholarly and activist critiques of neoliberal privatization have turned to the commons for redressing social and ecological injustices, Indigenous studies and decolonial scholars have problematized this concept and the implications of making sweeping claims about the commons. To date, major theorists of the commons have paid scant attention to the ambivalent role played by the commons in the colonial project. They have rarely engaged Indigenous perspectives on land tenure and modes of belonging to land as contributions for rethinking the commons. In the context of the US settler colonial state, for instance, claiming commons runs the risk of construing Indigenous claims to land and sovereignty as regressive and exclusionary. Chickasaw scholar Jodi Byrd (2011) makes this point in response to Nandita Sharma and Cynthia

Wright's (2008) proposal of instituting global commons as a key strategy of decolonization.[4] According to Byrd, the problem of reading the global commons as a generalized redistributive response to colonial theft is that it casts ongoing Indigenous struggles as an obstacle to a larger progressive project. The global commons are rendered, Byrd writes, "as the means to the end of oppression within the land that once did, but no longer can or should, belong to Indigenous peoples" (2011, 204).

In a similar vein, J. Kēhaulani Kauanui (2015) recalls the European origins of the commons and calls for "decolonial reflexivity" in relation to the concept (ibid., 3). She invites those struggling for a recovery of the commons to attend to the Indigenous difference to avoid strengthening the structures of settler colonialism that operate through the logic of the elimination of the Native (Wolfe 2006). To this end, it is crucial to resist "any conflation of 'the commons' with Indigenous communal lands, they are not one and the same" (Kauanui 2015, 6). This requires reckoning with the nexus between Whiteness and property (Harris 2003) in historical processes of racialized accumulation based on slavery, the seizure of Indigenous land, and the genocide of Native populations. Drawing on historical work by Allan Greer, Kauanui recalls the uneven encounter between Indigenous collective forms of land tenure and settler colonial uses of the commons in the New World. As they traveled across the Atlantic, the commons became a means of dispossession, entangled with European property claims over Indigenous land.[5] Yet, contemporary accounts of the commons often forget the fraught histories of the colonial commons and prefer to celebrate it as an inherently anti-colonial, resistant mode of living whose resurgence leads toward multiracial liberation. For Kauanui, Noam Chomsky offers an example of the conflation between struggles for the commons and Indigenous struggles. Indigenous activism in North America and beyond, he has argued, is a strenuous attempt to protect what is left of subsistence economies whose existence was once acknowledged in the Charter of the Forest, the companion document to the medieval Magna Carta. What Chomsky forgets, Kauanui points out, is that while the Magna Carta laid the foundations for the liberties of European individuals, those same liberties were asserted by settler colonialists in their proprietary claims over Indigenous land. For avoiding such omissions, it is vital to engage with Indigenous movements that refuse to disappear from their homelands (Simpson 2014; Estes 2019). Rather than using the language of the commons to explain them, it is crucial to attend to concepts and practices that connect Indigenous people to one another, to place and land, even as they do not easily align with non-Indigenous experiments with alternative futures.

Finding points of connection between a Marxist analysis of primitive accumulation and Indigenous struggles, the Yellowknives Dene political theorist Glen Coulthard invites us to reorient the understanding of the commons through decolonial lenses. Making decolonial commons, he suggests, is not just about access to land. It is about regenerating socio-natural relations, among people and between people and land, that centuries of colonization have attempted to suppress. This should not be confused with narratives of the ecological Indian that understands Indigenous people as custodians of healthy human-nature relations. A product of White imagination of the Indian capable of living a virtuous life in perfect balance with nature, the ecological Indian is a figure cast out of (Western) civilization and depicted as a victim of its excesses. In contrast, for Coulthard, regeneration refers to "the affirmative enactment of another modality of being, a different way of relating to and with the world" (2014, 169). This effort is carried out through Indigenous struggles for sovereignty and autonomy, including the Indigenous movement at Standing Rock that in 2017 opposed the construction of the Dakota Access Pipeline, which are grounded in the Native saying, "the land does not belong to us, we belong to the land." This is why, Coulthard maintains, reimagining the commons requires the commitment to supporting Indigenous sovereignty. It requires attention to ongoing forms of colonial dispossession, including the dispossession of Indigenous world-making practices and forms of knowledge that do not entail a binary distinction between active humans and inert resources (Todd 2015; TallBear 2017).

Native American arguments on accounting for Indigenous difference resonate, at least in part, with decolonial interventions around Indigenous movements in Latin America. Walter Mignolo, for instance, has warned against the universalizing deployments of the concept of the commons. He argues that the *common,* both in the imperial formulation of the British Commonwealth, and in the current leftist reworking of the term, emerged from the Euro-American epistemology and mode of existence. As such, it should be distinguished from the *communal,* a form of social organization that existed in precolonial times and continues to exert its influence in contemporary Indigenous society and politics. While the common is a leftist revolutionary project, one that has been reworked within Marxism, the communal is a decolonial project rooted in precolonial societies and in the Indigenous experience of five hundred centuries of coloniality. In this view, the Zapatista model of territorial governance provides a powerful instance of the reconfiguration of the communal rather than the commons. Confusing the communal with the commons, or worse subsuming the former into the latter, would be an unmistakable gesture of colonial appropriation (Mignolo 2011).

I take these interventions as necessary openings for reimagining the commons from decolonial perspectives. They provide crucial entry points for engaging the historical situatedness and positionality of the commons, and are useful for complicating ideas of the commons as a traveling project of liberation. I am indebted to decolonial critics who caution against the conflation of disparate stories of land dispossession in Europe and the colonial theft of Indigenous land. But I also want to distinguish the commons from European modernity, its racialized regimes of property, and constructions of (self-)possessive subjectivities. To be sure, the commons was born out of the ontotheological horizon of Christianity that has played a significant role in colonial project. However, it is also worth recalling that in medieval Europe, practices of common use were persecuted as heresy and largely expunged from that same horizon. As I recalled in Chapter 1, the thirteenth-century theological dispute over poverty involving the Franciscan order was resolved through the Roman Catholic Church's affirmation of man's natural ownership over the material world. The papacy's dismissal of common use has to be understood in the larger context of the emerging monetary economy and market forces that led to the enclosure of peasant land. This historical dynamic points to the tension between common use and the hegemonic forces that coalesced in the early modern formations of the individual, the state, and colonial capitalism. Instead of celebrating the commons as a universal paradigm of liberation or dismissing it because of its European origins, I want to draw attention to the paradox at the heart of the commons. On the one hand the commons has functioned as a counterpoint to European modernity, the heresy within, rather than as hegemonic formation. On the other hand, it has been implicated in the colonial project of dispossession and the production of exclusionary forms of humanism. Acknowledging this ambivalence, rather than affirming the commons as unifying paradigm connecting distinct struggles, may offer possibilities for encounters with decolonial perspectives.

The anthropologists Mario Blaser and Marisol de la Cadena (2018) approach the commons from the perspective of the *uncommons*, that is, heterogenous ontologies that might connect through struggles. Building on ethnographic work in Latin America, they maintain that Indigenous political movements express a political ontology that diverges from Western modernity in that it includes human beings as well as mountains, rocks, and water as part of place-based political collectives.[6] These movements challenge the hegemonic Western partition between humans who speak through politics and nature that speaks through science (de la Cadena 2010; Blaser 2013). By affirming the existence of disparate ontologies, they expose the Western notion of a single world as particular rather than universal (Blaser 2013; Escobar 2020). For Blaser and de la Cadena, uncommoning, the insistence on divergent modes of world-making, is the ground for creating commons, that is, alliances that arise from shared struggles. Commoning, in other words, does not consist in the composition of a common world but in the mutual transformation that might arise from the encounter between uncommons.

If decolonial scholars have engaged the dissonances between the commons and Indigenous political formations, scholars of the commons have been less adept at drawing on decolonial insights to explore how they might transform established understandings of commoning.[7] In this chapter I intend to move in this direction. I want to take seriously the argument, made by Indigenous scholars, that the universalist aspirations of the commons risk obliterating Indigenous demands for land, autonomy, and sovereignty. Still more, they risk ignoring modes of dwelling based on kin relations with nonhumans rather than the distinction between humans as agents of commoning and resources to be held collectively. Rather than calling for a generalized reappropriation of that which has been enclosed, I enter into conversation with decolonial insights to revisit struggles in Latin America—the Zapatista movement in Chiapas, Mexico, and the popular insurrection against the privatization of water in Cochabamba, Bolivia—that have often been interpreted through the category of the commons. But even as I problematize the universalist aspirations of the commons, I propose that situated projects of commoning and Indigenous communal modes of dwelling are potential allies in the creation of modes of existence alternative to advanced capitalism.

THE ZAPATISTA POLITICS OF THE LAND AND THE DEAD

Much has been written about the political impact of the Zapatista uprising across Latin America and beyond, its poetic language and the creative use of grassroots media, its impact on alter-globalization movements and other collective practices pursuing self-determination in the face of neoliberal dispossession. The ongoing project of autonomy from the state and the market set in motion in Chiapas, Mexico, in 1994, has been often interpreted through the category of the commons. The same concept has been used to analyze social movements in Bolivia that in the early 2000s paved the way for the pink tide government of Evo Morales. More recently, the commons have been evoked as an alternative to the intensification of extractive economies focused on export-oriented, large-scale projects based on the development of mining and hydrocarbon industries across Latin America (Svampa 2019). Complicating these interpretations, this chapter shows how political movements in Southern Mexico and Bolivia put pressure on contemporary Western understandings of the commons as collective resources or the product of human labor. These

Crossing the Storm: Encountering Decolonial Perspectives 145

political movements have brought to the fore a distinctive political ontology that, unlike the one developed throughout European modernity, involves a range of negotiations among human and nonhuman beings. Their enactment of ontological difference has become incompatible with the intensification of extractive economies.

The emergence of the Zapatistas out of the Lacandón Jungle in 1994 played an important role in revitalizing the political imaginary of the commons. Arising from the encounter between Indigenous groups and a small cadre of mestizos Marxist revolutionaries (Khasnabish 2008), the insurgency began the same day that the North American Free Trade Agreement (NAFTA), went into effect. One of the prerequisites for Mexico's participation in trade liberalization was the revision of Article 27 of the constitution that, following the 1910 Mexican Revolution, allowed small farmers and Indigenous communities to access the *ejidos,* communal land and water that comprised over 50 percent of Mexico's croplands and 80 percent of its forestland. On the first of January in 1994, the Zapatista army (EZLN), mainly composed of Indigenous Maya wearing balaclavas, seized six towns in the southern state of Chiapas, home to a large percentage of Tzeltal, Tojolabal, Tzotzil, and Chol populations. A young Tojolabal woman, whose *nom de guerre* was Major Ana Maria, led the takeover of the key city of San Cristobal de las Casas.[8] Major Ana Maria's role was not exceptional in the EZLN. Women made up from 30 to 40 percent of the military ranks of the Zapatista army. In 1994 the movement made public the Women's Revolutionary Law that countered masculine hegemonic practices within the postcolonial state by demanding women's full participation in the revolutionary struggle and in the governance of the community, control over reproduction, and access to education and health. I will return to Ana Maria in a moment. As the gun-shaped chunks of wood carried by the insurgents in the early days of the rebellion suggested, the Zapatistas have never been primarily a military force. Their goal has never been seizing state power in the manner of traditional Marxist-Leninist revolutionaries. Although the military structure of the EZLN is still in place, the movement has increasingly distanced itself from the vanguardist guerrilla model. Its central focus has been the creation of autonomous municipalities in recuperated land (the *caracoles*) seen as building blocks of a form of autonomous governance capable of eroding Mexican state sovereignty from within (Reyes and Kaufman 2015).

For Midnight Notes, a US-based research collective that in the 1990s identified continuous dispossession as the fundamental *modus operandi* of capitalism, the Zapatista upheaval was a powerful alternative to neoliberal new enclosures. The insurgency in Chiapas has been read as the reclamation of commons against the ongoing capitalist privatization of land and resources. By experimenting with autonomous governance, distinct from the Mexican state and the market, the Zapatistas have been doing more than resisting neoliberal regimes of accumulation. For Midnight Notes, they have made visible the commons as mode of being together that Marxist orthodoxy relegated to the past. Still more, the Zapatistas have demonstrated that the commons can function as a connector among struggles, "a place where Marxism, ecology, Indigenous, and antislavery struggles meet" (Caffentzis and Neill 2004, 63). Gustavo Esteva, a Mexican activist-intellectual who worked closely with the Zapatistas, makes a similar point about how the insurgents have sought to "reclaim their commons." But he also adds that their project is to "regenerate their own forms of governance and their own art of living and dying" (Esteva 2004, 10). The point about reclaiming commons *and* a particular art of living and dying deserves further exploration. If, as Walter Mignolo contends, for the Zapatista revolution to succeed "the Marxist cosmology needed to be infected by Indigenous cosmology" (2011, 219), what are the consequences of such an encounter for the politics of the commons? In order to address this question, I turn to the role of the land, the mountains, and the dead in Zapatista politics.

In their discussion of Subcomandante Marcos's writings, political theorists Margaret Kohn and Kelly McBride (2011) argue that by insisting on land as the precondition for the survival and flourishing of Indigenous identity, the Zapatistas brought together claims of identity recognition, economic redistribution, and political self-determination that are usually distinct. While I follow Kohn and McBride's turn to the relevance of the land in Zapatista politics, I explore a different dimension of land, one that builds on Blaser and de la Cadena's argument that current Indigenous mobilizations in Latin America challenge Western hegemonic conceptions of politics grounded in the ontological distinction between humans and nature. Many commentators understand Latin American political movements of the 1990s and the early 2000s as responses to the neoliberal extraction of resources and as socio-environmental mobilizations protecting territories that are crucial for the subsistence of Indigenous populations. Alternatively, or in combination with such interpretations, these movements are seen as articulating ethnic claims connected to religious and cultural beliefs. According to de la Cadena, however, the lenses of political economy and cultural difference are insufficient to understand the uprisings. Indigenous movements, she argues, have been summoning into the political field "sentient entities whose material existence—and that of the world to which they belong—is currently threatened by the neoliberal wedding of capital and the state" (de la Cadena 2010, 342). What is at stake is not simply the affirmation of Indigenous "beliefs" but a disagreement between worlds regarding the ontological makeup of politics, that is, who and what counts as a political actor or concern.

Inspired by this argument, I consider the land and the dead as an integral part of the Zapatista experiment in autonomous governance. Looking at the early years of the insurrection, I examine two speeches delivered by Zapatistas spokespersons. The first one was given in Mexico City by the Subcomandante Marcos. In 2001, wearing his signature ski mask, he spoke before 250,000 people gathered for the march for Indigenous Dignity in the Zocalo, the square where colonial buildings stand next to Aztec ruins. This intervention in a public space associated with power and protest, wove together modernity and its alternatives in a rich tapestry of images and concepts. It reclaimed liberty, justice, and rights, that which Western modernity has affirmed as universal but whose access has been systematically denied to Indigenous people. But the speech also exceeded the political vocabulary of modernity by addressing Indigenous relations to land.

One of the central motifs of Marcos's speech was the mirror. Addressing the crowd gathered at the Zocalo with his back toward the presidential palace, he claimed:

> We are a mirror. We are here in order to see each other and to show each other, so may look upon us, so may look at yourself, so that the other looks in our looking. We are here and we are a mirror. Not reality but merely its reflection. Not light, but merely a glimmer. Not path, but merely a few steps. Not guide, but merely one of the many directions which lead to tomorrow.[9]

Even as the claim of constitutional recognition of Indigenous rights and culture is explicitly articulated in the speech, the Zapatistas are not just demanding that the Mexican authorities recognize them as counterpart. They are also disrupting the traditional relation of recognition by which, to put in Hegelian terms, the slave can only acquire subjectivity when they are recognized by the master. The mirror stands for the masked, faceless Zapatistas, the insurgents who have turned the colonial deprivation of individuality into a weapon of collective transformation. At the same time, the mirror faces Indigenous people who can see a trace of themselves in it, a fragment of collective life autonomous from the postcolonial state. Simultaneously, in a powerful dismissal of political vanguardism, the mirror also shows the Zapatistas as "reflected light," something whose power is amplified by a larger context of struggles. In shifting from recognition to autonomy (Coulthard 2014; Reyes and Kaufman 2015) the Zapatistas create a political alternative to modernity. But there is more.

The speech at the Zocalo affirms that the land is involved in the insurgency in ways that would be unthinkable from the modern perspective. In Marcos's words, "We are rebels because the land rebels if someone is selling and buying it as if the land did not exist, as if the color of the earth did not exist." The Zapatista rebellion expresses more than a revolt for reclaiming the land seized throughout the ongoing conquest of Indigenous territory. It expresses the insurgence of communities for whom the land is not a resource at the disposal of human inhabitants but a landscape populated by entities with whom humans have to negotiate. When Marcos refers to the insurgency *of* the land, he is evoking a cosmology with profound implication on political life. As Mihalis Mentinis elaborates, the struggle of the Zapatistas is dynamically informed by a cosmology in which being human means depending on an animal co-essence (*nagual*) that defines one's power to act. For the insurgents, "Mountains and forests are alive, they have their *ajaw* (Tzeltal for 'guardian of the mountain'), each of them having specific powers and their distinct character" (Mentinis 2006, 154).

In order to further elaborate this point, I now turn to the speech delivered in August 1996 by the insurgent leader Major Ana Maria to welcome thousands of activists gathered in the Zapatista community of Oventic for the Intercontinental Encounter for Humanity and Against Neoliberalism. Major Ana Maria (1996) spoke an "improper" Spanish, a language hybridized with elements of Tojolabal, her Indigenous language:

> For Power, that Power now clothing itself all across the world with the name of "neoliberalism," we did not count, we did not produce, we did not buy, we did not sell. We were a useless number in the accounts of big capital. Then we went to the mountains searching for the good and to see if we could find alleviation for our pain, of being forgotten stones and plants. Here, in the mountains of the Mexican Southeast, our dead live. Our dead, who live in the mountains, know many things. Their death spoke to us and we listened. . . . Little boxes that speak told us another story that comes from yesterday and points towards tomorrow. The mountains spoke to us; the *macehualtin*, those who are ordinary and common people, and we the simple people, as we are called by the powerful. . . . Behind our black face, behind our armed voice, behind our unspeakable name, behind the we that you see, behind us we are (at) you (*Detras de nosotros estamos ustedes*).

In an acute analysis of the concluding sentence of the speech ("*Detras de nosotros estamos ustedes*") Walter Mignolo and Freya Schiwy (2002) point out that what is most striking in Major Ana Maria's words is the displacement of the subject-object correlation, the I and the you, which characterizes the pronominal structure of European languages but not Amerindian languages. Descriptions of actions in Indo-European languages are based on the presence of an acting subject, a verb, and an inactive recipient of the action. This is what we find, for example, in the sentence "I told you," where the "you" is presented as essentially passive with respect to the speaking subject. In contrast, Tojolabal speakers say something like "I told, you heard," to describing a bidirectional, intersubjective process (Lenkersdorf 1996). In the last sentence of the quote above, Major Ana Maria's native language, Tojolabal, infiltrated Spanish in a peculiar way. Through the misuse of the word *estamos* rather than *somos*, she confounded the boundaries between the one who speaks, the Indigenous woman acting as spokesperson of her political community, and those who were listening, many of whom were non-Indigenous activists. Such linguistic transgression offers a glimpse into a cosmology out of which the grammar emerges, one in which "persons, living systems, and nature are not objects but subjects" (Mignolo and Schiwy 2002, 8).

In Major Ana Maria's speech, the Mayan cosmology is also present at another level. She referred to a conception of time that moves in cycles rather than linearly ("another story that comes from yesterday and points towards tomorrow"). Insofar as the Zapatistas see the past as the point from which the dead speak to the future, they invest in the revolutionary possibility of a world turned upside down. The Indigenous cosmovision also emerges in the references to mountains inhabited by ancestors who speak and pose problems to which the insurgents have to respond. The neoliberal Mexican state relegated Indigenous people to the status of "forgotten stones and plants," that is, less-than-human beings assimilated to nature, and therefore destined to be ruled. Something quite different happens in the mountains of the Mexican southeast where mountains and the dead enter into the composition of a political community that not only challenges the dehumanization of Indigenous people but also the partition between society and nature. Interpreting the Zapatista insurgency simply through the category of the commons, as the reclamation of land and resources against ongoing processes of enclosure, misses the potential of the encounter between Marxist guerrillas and Mayan communities. What has flourished in Chiapas, albeit intermittently and under precarious conditions, is not just an insurgency against new rounds of capitalist enclosures. Although the process of land recuperation and the possibility of collectively working it to sustain the Zapatista communities has been central, what has emerged is also an insurgency of Indigenous political ontology, the reactivation and reinvention of modes of living together on ancestral land that puts pressure on the assumptions about the inert earth that supports the architecture of

Western politics. This insurgent ontology openly challenges notions of land as a commodity to be incorporated into the global expansion of capitalism but it also complicates understandings of the commons as shared management of natural resources. Communal land does more than just provide the Zapatistas with the indispensable sources of subsistence to build a new society based on gender equality, shelter, health, and education for all. Land, mountains, forests, and the dead are integral actors in practices of insurgent world-making.

EARTH-BEINGS IN THE BOLIVIAN WATER COMMONS

In the early 2000s, massive uprisings against the privatization of water in Cochabamba, Bolivia, ushered in a profound political transformation of the Andean country. The water-war, followed by widespread roadblocks organized by Indigenous peasants and coca growers, is often described by scholars and activists as a struggle for the autonomous management of common resources. In *Commonwealth*, Michael Hardt and Antonio Negri (2009) argue that Bolivian social movements were based on the common in two intertwined senses: They claimed access to resources and organized themselves through preexisting forms of Indigenous communal self-government. Ugo Mattei, an Italian legal scholar of the commons, writes that the Bolivian constitution, introduced in 2009 largely as a result of social struggles, is "the most advanced juridical model elaborating the concept of common good available to humanity if only it would set aside Western arrogance and rethink its model of development" (2011, 22). In what follows, I bring to the fore aspects of these struggles that are often overlooked by most theorists of the commons. Specifically, I pay attention to the force of water in the political collectives that mobilized in Cochabamba.

Bolivia, the Andean-Amazonian territory with a large Indigenous population, mainly Aymara and Quechua, has been a laboratory for neoliberal policies since the mid-1980s. In the late 1990s, the World Bank negotiated with the Bolivian government the privatization of the public water system in Cochabamba, the country's third-largest city, in exchange for $600 million in debt relief (Hindery 2013). After the public water system was sold to the consortium Aguas de Tunari, controlled by the multinational corporation Bechtel, water rates soared and local communities risked losing access to sources of water that autonomous neighborhood groups and committees had long managed without state concessions. The committees, mainly established at the city's edges, had devised a variety of ways to supply drinking water to local communities, including rainwater collection systems, the drilling of wells, and the use of private delivery trucks. Organized through a collective decision-making process, they also determined whether and how to connect to the municipal water supply system. According to the Bolivian water activist Marcela Olivera, the water committees were (and still are) the expression of a process of autonomy "based on practices recognized neither by the state nor by the international community—and that need no recognition in order to exist" (2015, 88).

In response to privatization, the people in Cochabamba organized through the Coordinadora de Defensa del Agua y de la Vida (Coalition for the Defense of Water and Life) that included water committees, farmers, Quechua Indigenous groups, factory workers, and women's organizations. The waterwar quickly spread to the western Andes of Bolivia. Aymara groups in Oruro and La Paz opposed a bill in congress that would have privatized water in the region on the grounds that it violated Indigenous understandings of water (Webber 2011). A wave of strikes and road blockades shut down large areas of the country and ultimately forced Aguas de Tunari out of Cochabamba. The water-war was a catalyst for a period of prolonged political unrest. In 2003, protests against the privatization of natural gas raged in the Indigenous city of El Alto, a satellite of the capital city of La Paz, and led to the resignation of Bolivian President Gonzalo Sánchez de Lozada. Simultaneously, a movement of unionized coca growers contested the US-led drug policies in the Andes and eventually catapulted the Indigenous cocalero Evo Morales into the presidency. Until 2020, when it was overthrown by right-wing forces, Morales's government enacted highly contradictory policies. While it recognized the Indigenous Pachamama as subject of rights and introduced anti-poverty measures, Morales's administration aggressively pursued the expansion of state-controlled extractive industries. Supported through a network of interdependence with foreign capital, large-scale hydrocarbon and mining projects have transformed Indigenous landscapes. This caused a profound rift between the government and the social movements that brought Morales to power (Webber 2011; Rivera Cusicanqui 2015; Fabricant and Gustafson 2016; Tola 2018).

Although the Bolivia water-war involved a multitude of organizations, it was largely animated by Indigenous groups and conveyed various elements of Indigenous cosmovisions. The evocations of Pachamama and other figures of Andean cosmology during the protests are significant in this sense. In Cochabamba, for example, campaign posters referred to the role of Andean deities in the resistance to neoliberal privatization: "Pachamama, Wiracocha, and Tata Dios gave (water) for us to live, not to do business with" (Andolina et al. 2009, 144). The Aymara sociologist Pablo Mamani Ramirez helps us to makes sense of this slogan. He writes, "in the logic of the *ayllus*, water cannot be bought or sold, or subjected to market logic because water is a vital part of life: it is the blood of the Pachamama" (2004, 81). This means that in the *ayllu*, a precolonial form of collective life which is still practiced in the Andes, water constitutes an expression of Pachamama, a being that participates in the life of Indigenous communities.

In the Andean cosmology, Pachamama is the matrix of life that includes the earth. Despite being disqualified by the scientific paradigm of colonial modernity, it remains a central figure in Indigenous cosmologies that still shape much of Andean life. Pachamama has been described as the earth's generative powers (Silverblatt 1987), and as the vitality that animates the earth at a particular moment in time (Allen 2002). Although it is usually translated as Earth-Mother or World-Mother, it was originally deprived of maternal qualities, least of all those associated with the chastity and benevolence of the Virgin Mary. After European colonization, however, the Andean deity was progressively assimilated to the Virgin and rendered as a nurturing mother (Tola 2018). If Pachamama embodies the earth's generative vitality, water constitutes one of its multiple expressions, it is "the blood of Pachamama" that allows Indigenous communities to flourish.

Mamani Ramirez's reading of water in relation to the *ayllu* and Pachamama differs from that of Jeffrey Webber, author of various studies of Bolivian political movements. A Marxist scholar, Webber interprets the *ayllu* essentially through categories of socioeconomic organization. He describes it as a precolonial formation that still informs life in the Andes through the communal control over Indigenous territory and land. This definition brings the *ayllu* in close proximity to the commons as both seem to designate the communal use of land and natural resources distinct from state and private property. For Webber, the Bolivian uprisings expressed "racialized peasant class-struggle" (Webber 2011, 172). They were born out of the explosive convergence between long-standing Indigenous struggles against state racism and the opposition to neoliberal policies threatening the communal management of water and land. This account connects the opposition to neoliberal privatization to Indigenous conceptions of "natural resources" but, concerned as it is with illustrating the development of "racialized peasant class-struggle," it does not consider the extent to which these conceptions informed political movements.

Like Mamani Ramirez, Aymara intellectual Marcelo Fernandez Osco's interpretation of the *ayllu* is informed by Indigenous cosmologies. He explains that the *ayllu* "keeps order by maintaining an understanding of the sacred character of everything. This sacral sense charges the notion of life with resonances that expand beyond humans to include a multiplicity of life forms that are not considered in asymmetrical or objectivist terms" (Osco 2010, 30). In light of this definition, it is easier to grasp why, in the logic of the *ayllu*, the privatization of water is problematic. Water is not simply a resource but part of a collectivity whose persistence and possibility of flourishing is threatened by neoliberal development. Marisol de la Cadena further clarifies the logic of the *ayllu* when she defines it as "the socio-natural collective of *tirakuna* (the sentient beings made of earth and water) as well as of humans, animals and plants, inherently connected to each other, so pervasively that nobody within it escapes such relations, for it is such relation, the *ayllu*, that makes the place and the persons who live in it" (de la Cadena 2013, 59). The *ayllu* is made up of the exchanges among earth-beings and other actors. Together, they form a place-based collective that in many Andean contexts assumes insurgent valences when confronted with neoliberal privatization. Framing the *ayllu* as a socio-natural organization comprising other-than-human expressions of Pachamama opens up space for considering the generative friction between prevalent Western approaches to the commons and Indigenous reconfigurations of the communal. Instead, interpreting political movements in Bolivia as the product of "racialized peasant class struggle" risks obliterating significant specificities of these struggles.

The relationship between the commons and Indigenous ontologies is productively articulated by Marcela Olivera, who was the international liaison for the Coordinadora de Defensa del Agua y de la Vida in Cochabamba and went on to develop the water justice network Red VIDA. The author of several essays on the water-war and a bridge figure among activist worlds between the United States and Bolivia, Olivera points out that in contemporary Cochabamba, hundreds of water committees have reconfigured communal practices that can be traced back to precolonial times and adapted them to contemporary urban struggles for water autonomy. In the essay "Working with the Commons," she writes, "Andean cosmologies largely consider water as both a living being and divine being. Water in the Andes is the basis of reciprocity and complementarity, it helps to resolve problems and to establish relationships. Water is everyone's and no one's. It is the element that helps nature create, transform life, and permit social reproduction" (Olivera 2019). Here Olivera renders the lived experience of struggle in Cochabamba using both the language of the commons and that of Indigenous cosmovisions. She describes the autonomous water supply systems in Cochabamba as commons while at the same time indicating that water is a crucial element in the rich fabric of socio-natural relations that make up life in the Andes.

As an architect of activist networks across the Americas, Olivera highlights the Bolivian water-war as a site of encounter between the commons and the Indigenous relations to land. Rather than describing autonomous networks of water distribution as an expression of either one of these modes of collective life, she brings them together so that they become inseparable and yet subtly distinct. In using the category of the commons to describe struggles for water in Bolivia, one that is perhaps more easily accessible to transnational activist networks, she is careful in not obscuring what makes the commons different from Indigenous cosmovisions. Weaving together the two rubrics suggests that autonomous water committees in the Andes are not just claiming common access to resources and organizing themselves through preexisting forms of Indigenous self-government. They are also affirming that what Western political ontology presents as "natural resources," are beings that participate in political collectives. These struggles, in other words, blur the boundaries between political agents and things that have to be managed in the exercise of political power.

CONCLUSION

Between October and December 2023, while organized crime groups clashed in Chiapas to gain control of the Mexican border with Guatemala, the Zapatistas released twenty communiqués announcing major changes in their self-governance structure. These are meant as ways for the insurgent Indigenous movement to defend itself from the lethal cruelty of the cartels but they are also a response to a broader existential threat, what the Zapatistas call "the storm" (Deslandes 2024; Zibechi 2024). This term indexes a combination of climate and environmental disasters, the intensified violence of war, paramilitary gang activities, and extractive industries. Embracing a long-term, seven-generation outlook, the insurgents have adopted several measures to cross the storm, to survive the darkness and reach the morning, when, they write, a young girl, embodying a new generation of Mayas in relation to land, will learn "that being free is also being responsible for that freedom."[10] To this end, the thirty autonomous municipalities and the twelve Good Government Juntas, key elements of the Indigenous self-governance, have been dissolved and replaced by one thousand decision-making bodies that will confederate to ensure connections between local communities. This process of decentralization aims to protect localities while avoiding isolation in the face of the storm. The twentieth and last communiqué explains that central to the new direction are forms of land use comprising personal and family property coexisting alongside *el común* ("the common"), that is, land that does not belong to anyone, land "without papers" that can be worked collectively as the material basis for the reproduction of Zapatista autonomy. A small part of this nonproperty will be open to people coming from other geographies so that they can learn from the Zapatistas how to work the land and respect it, while, in turn, offering their knowledge and ways of working the land.[11]

No one knows how the Zapatista's seven-generation project will unfold. But it is significant that they envision *el común* as a practice for recrafting their conditions of existence and enable life in the present and in times to come. For my part, I look at the patches of land shared with activists from other geographies as spaces for reproducing life while also experiencing the discontinuities as well as the possibilities of alliance between ways of making commons. In doing so, I join Mario Blaser and Marisol de la Cadena in reframing the commons as a "continuous achievement, an event whose vocation is not to be final because it remembers that the uncommons is its constant starting point" (Blaser and de la Cadena 2018, 19). Even as the Zapatistas center localities, they maintain openings for renewing onto-epistemological and political encounters patiently forged over many years of struggles. Making commons might include disagreements (including those about the status of nonhuman beings) but in addition to reproducing the autonomy of specific worlds, it is also a practice of relation across distinct modes of making insurgent worlds.

This relational practice is at the core of this chapter. Focusing on moments of uneasy encounters between movements from the Northern and Southern Hemispheres, it seeks to illuminate the commons and Indigenous insurgent political ontologies as distinct and yet conjoined in multi-sited projects of crossing a storm that lands differently in the uneven landscapes of extractivism. It is my contention that both the commons and Indigenous socio-natural organizations diverge, although in different forms and degrees, from the dominant Western political ontology that assumes the individual man as the organizing principle of political life: A self-sufficient and self-possessing person separated from other living beings through the hierarchical ordering of sexual difference, race, and species (Cohen 2009; 2013). To be sure, there are important differences between them. Indigenous struggles are not just resisting new enclosures, they are challenging colonial capitalist formations that turn land into a mere site of value extraction. In the process, these movements are reenacting relations to land that encompass people, animals, plants, and rocks (Coulthard 2010). Contemporary articulations of the commons in political theory and activism powerfully reactivate the heretical dimension of the commons as a mode of using without appropriating the world. But in claiming the commons as alternative to heterogenous modes of capitalist accumulation, they often cast minerals, land, water, animals, and plants as resources transformed through human cooperation. Attending simultaneously to the proximities and divergences between the commons and Indigenous movements is crucial for creating spaces of alliance directed at challenging the violence of varied forms of extractivism. Such alliances, however, can only arise from a radical reassessment of the common's aspirations to universalism. This means not just dealing with the existence of diverse instances of commoning that would need to find articulation into a larger political project. Rather, it entails doing away with the investment in the "one-world world" (Law 2015; Blaser 2025), that is, the assumption that there is just one humankind and one nature, to better appreciate what the Zapatistas call "a world where many worlds fit."

EPILOGUE: ON INSTITUTING

To conclude this book, I would like to return to the place evoked in the opening pages: The permanently inundated North American city imagined by Kim Stanley Robinson in *New York 2140*. In this sunken urban sprawl, capitalism seeks new opportunities for profit. While the superrich have fled the flooded areas, diverse collectives have assembled in the intertidal zones of Midtown and Downtown Manhattan creating unforeseen political possibilities. Even as this speculative fiction exposes in great detail the depth and scale of environmental crises and climate inequalities brought about by economic and financial systems, its main focus is the convivial modes of living that exist in spite of and against capitalism. The novel (spoiler alert) culminates in a jubilant moment: Life, the narrating voice claims, "is going to explode the enclosures and bring back the commons" (Robinson 2017, 243). A new storm prompts the dispossessed to move Uptown and set up encampments in Central Park. Riots spread within and beyond New York and the United States. People refuse to pay rent, mortgages, and loans in an "everybody strike" leading to a massive financial crisis. Eventually, the strike's demands translate into the election of progressive officials into Congress, in measures supporting labor and protecting the environment and, strikingly, in the nationalization of banks. These policies have ripple effects at the global level as other countries move in a similar direction.[1] The narrator warns that these are "transient political accomplishments" with "no guarantee of permanence" because "every moment is a wicked struggle of political forces" (ibid., 453). Still, the novel provides an optimistic outlook in which the state and even financial actors play a positive role in destabilizing the neoliberal order.

Moving from this imaginary timespace back to Rome, the Southern European city where part of the research for this book was conducted, my outlook is decidedly more somber, affected as it is by a dire political climate for many groups, locally and globally. The years and months leading to the completion of *Resurgent Commons* witnessed the rise of authoritarian variants of neoliberalism and the proliferation of elusive masculinist microfascisms.[2] These developments have fueled the exacerbation of economic inequalities; the stubborn attachments to fossil fuels and a renewed mastery of nature; the normalization of racial violence and lethal border containment in Europe and the United States; prolonged colonial warfare, genocide, and environmental devastation in Palestine; and gender-normative discourses and policies targeting trans, queer, and women's self-determination.[3] In Italy, these reactive formations have been embodied by Giorgia Meloni, the leader of a far-right party with neofascist roots. Her climb to power came to be defined by the motto, "I am Giorgia, I am a mother, I am a Christian," an assertion of reactionary White femininity pragmatically wedded to neoliberal policies that exacerbate poverty and precarity while blatantly ignoring the climate crises.[4] At this juncture, identifying paths of situated yet expansive collective transformation is a difficult but crucial task. Resurgent commons, in my view, provide openings in this direction. Affected by specific political and ecological conditions, never identical to themselves, resurgent commons continue to emerge in myriad locales. Often, although not always, they trace liberatory becomings through and against varied enclosures, persisting in the effort of crafting futurities.

In Robinson's novel "the comedy of the commons" constitutes the brewing ground for ushering in forms of political representation that finally meet the magnitude of the socio-ecological crisis. It is not my intention to easily dismiss the investment in state politics, particularly when, as in Robinson's case, it is coupled with the emphasis on the role of grassroots politics in its democratization.[5] But my investigation points to different, if overlapping, priorities. Rather than probing the ways in which commons can translate and scale up into state policies, I am interested in their *persistent resurgence*, in their capacity to put themselves into question while expressing an instituting praxis that exists in conflictual autonomy from the state and strives to create socio-ecological relations alternative to exploitation and extraction. In thinking resurgent commons and institutions, I draw on the instituting practices enacted in the Southern European city of Rome by the projects analyzed in the previous chapters, including the transfeminist anti-violence shelter and political hub Lucha y Siesta (Chapter 4) and the Bullicante Lake, a more-than-human assemblage catalyzed by the force of a body of water (Chapter 5). These experiences might be partial and circumscribed in space and time, but they provide insights about the instituting potential of resurgent commons in contexts of the neglect, dysfunction, and hostility of state institutions. What is the relationship between resurgent modes of commoning and institutions? To what extent is the political praxis of instituting commons capable of bypassing or transforming established institutions, including state institutions?

As I noted in Chapters 4 and 5, resurgent commons in Southern Europe do not place state institutions on the same plane as market actors but demand state interventions and welfare provisions in a range of matters and at various levels, from funding for organizations supporting women fleeing patriarchal violence, to environmental protection and the maintenance of public spaces against privatization. Acutely aware of power maneuverings, they deploy contentious actions that challenge established institutions in a number of ways. Consider for instance when in early summer 2025 activists gathered at the Bullicante Lake dressed up like trees and birds and took to the streets, walking or biking for about three miles toward the Campidoglio, the seat of Rome's administration. This "walking forest," a citywide ecological alliance in part inspired by Bullicante Lake's commons, called on the mayor and the municipality to protect more-than-human ecologies within the city and open up to participatory decision-making in matters of urban environmental policies. This action advanced clear demands to political authorities while translating convivial ways of inhabiting the earth into a performative protest. But the ecopolitics of the commons is more about affirmation than recognition. It entails, first, a refusal to be governed, and second, a praxis of self-governance grounded in the rhythms and desires of collectives, places, and territories, in archives of knowledges from below, memories of combats, and efforts to cultivate connections with other struggles. This means prioritizing processes of everyday autonomous instituting while also seeking to transform established state institutions that often stand in the way of resurgent commons, regulating access to infrastructures that are key for communal organizing and routinely advancing private interests in the name of the public good.[6] From the vantage of resurgent commons, it makes sense what Michele Lancione notes in discussing the politics of home and homelessness: The state "should be no more and no less than the facilitator for multiple plans to emerge, not the one doing the planning" (2024, 205).

Building on Gilles Deleuze and Antonio Negri, the Rome-based activist scholar Francesco Raparelli (2021) proposes to rethink institutions not from the transcendent perspective of state sovereignty but from the perspective of immanent imagination and embodied social intelligence.[7] Alternative institutions, he writes, "display the inventive and public character of our species, naturally artificial and historical" (ibid., 112). This means that they are provisional, susceptible to continuous restructuring, and attentive to how power circulates. The emphasis here is on the creative capacities of the human species arising from social struggles. This is a compelling proposition, but this book shows that processes of instituting otherwise depend on the organized force of bodies in relation with places that do more than just contain them. Place-based attachments, the affects, meanings, and relations emerging over time in the interplay between bodies and spaces, are central to the instituting praxis of resurgent commons, one that strives to persist over time while also remaining in flux.

In the process of writing this book, I have returned many times to Anna Tsing's conceptualization of the commons in *The Mushroom at the End of the World* (2015). Tsing traces the matsutake commodity chain from North America and China to Japan, noticing how practices of matsutake picking are part of "latent commons," that is, "entanglements that might be mobilized in common cause" (ibid., 135). Latent commons, Tsing maintains, are ubiquitous but difficult to notice. In the context of the Pacific Northwest, for example, this term can be used to capture ways of composing loose communities in forests that are often experienced as commons as well as attempts to organize precarious pickers, many of whom are war refugees, undocumented, and racialized migrants. But in the case of matsutake, there are also the underground commons formed through the exchanges that allow human-disturbed forests to become fungi-forming sites and, in turn, allow matsutake to provide trees with nutrients. Latent commons take shape in law's interstices, proliferating informally under the radar of contracts formalizing property arrangements. In this sense, Tsing points out, "latent commons do not institutionalize well. Attempts to turn the commons into policies are commendably brave, but they don't capture the effervescence of the latent commons" (ibid., 255).

Tsing's invitation to develop "many kinds of alertness" (Tsing 2015, 254), to notice the more-than-human entities and beings contributing to latent commons has been invaluable to my project. Yet, as other scholars have noted, the notion of latent commons stresses collaborative survival in capitalist ruins rather than collective flourishing. These are not the same thing. Indeed, surviving amid capitalist ruins can be exhausting and unsustainable over time even when collaborations are involved. So, while this concept gestures toward the escape from capitalist capture and the possibilities of struggling in and across varied conditions of accumulation, it leaves open the difficult question of how to organize and give consistency and continuity to processes of decommodification in contexts of capitalist extraction. I do concur with the argument that commons do not institutionalize well and would add that efforts to distill and replicate design principles across commons, even when well intentioned, miss important specificities and even risk favoring their incorporation by state and market formations. At the same time, resurgent commons in Rome suggest that self-organization can lead to an autonomous instituting praxis aimed at broadening decommodification while retaining effervescent qualities in unfavorable conditions. At times, this praxis requires tense negotiations with political authorities, uneasy interlocutions, and demands for policy change to facilitate the persistence and broadening of commons.

Drawing on feminist environmental humanities and political ecologies and in conversation with decolonial perspectives, this book has sought to unpack several crucial aspects of the commons. It has traced underexplored and ambivalent histories of this concept (Chapters 1 and 2) thus inviting a reconsideration of scholarly and activist claims that insist on its inherently emancipatory valences (Chapter 6). Deploying the term uncommoning (Blaser and de la Cadena 2018), it points to problematic universal aspirations of the commons (Chapters 2, 3, and 6), calling for sensitivity toward differences and divergences between distinct, and at times illegible, ways of making worlds. It has been my contention that there is no one way of commoning and no one-size-fits-all unifying theory explaining evolving dynamics of enclosures and resistance in the landscapes of global capitalism. There are, however, multiple rhythms and materialities of resurgent commons that are worth paying attention to in the effort of ensuring the flourishing of all in times of accelerating socio-ecological injustices. As this book demonstrates, resurgent commons come into being through the interplay of bodies and places, human and other-than-human beings, providing measures of protection and openings to cut through the violence of exploitation and extraction in which state institutions are deeply implicated.

Again, there is nothing romantic in reactivating and instituting commons amid ruins.[8] In Rome, where ruins hark back to past empires and signal persistent differentials of power and privilege, commoners are confronted with legal frameworks designed to uphold private property. They are confronted with structural inequalities along the fault lines of gender, race, class, and nationality. Many, although not all, face housing and income precarity, exhaustion, and burn out, the police, bureaucracy, and eviction orders. And yet, there is joy in commoning. This praxis, when animated by a critical awareness about its own troubles and contradictions, remains crucial for inhabiting the earth otherwise and generating worlds that are worth living in.

ACKNOWLEDGMENTS

My deepest thanks to the many people and places composing transformative commons in Rome, the European South, and beyond: the political laboratory and feminist home Lucha Y Siesta; the human and nonhumans of the Bullicante Lake; the Non Una di Meno transfeminist assembly in Rome. This book would not exist without them.

In the process of conceiving and writing *Resurgent Commons* I enjoyed tremendous support from many colleagues, friends, and students. Huge gratitude goes to collaborators and colleagues who influenced the trajectory of this manuscript or gave feedback on portions of it: Maddalena Fragnito for being my coconspirator in thinking about the ambiguities of care; Tania Rispoli for joining forces in writing about feminist infrastructures and socio-ecological reproduction and the editorial director of *Feminist Studies* Ashwini Tambe for suggesting that we work together; Ugo Rossi for coauthoring a chapter about "the common"; Bruce Braun and Sarah Nelson for pushing me to articulate my thoughts on autonomist Marxism in the Anthropocene as editors of a special issue of *South Atlantic Quarterly*; Marco Armiero, Stefania Barca, and Roberta Biasillo, inspiring scholars of radical environmental humanities, for their feedback on more-than-human commons in Italy; the fantastic partners of the research team on Intersectional Political Ecologies of the Commons based at the University of Accra and the University of Bern: Akosua Darkwa, Deniz Ay, Pambana Bassett, and Leandra Choffat; the Politics Ontology Ecology (POE) research network and especially Maura Benegiamo, Emanuele Leonardi, and Luigi Pellizzoni for creating a welcoming space for many convivial conversations on political ecology.

This book was many years in the making. The early stages of research were conducted while I was a PhD student in Women's, Gender, and Sexuality Studies at Rutgers University. My advisors, Ed Cohen and Elizabeth Grosz, provided unwavering guidance, intellectual provocations, and friendship. They encouraged me to think outside the box of academic disciplines and have fun in the process. It was a great honor having Ethel Brooks and Michael Hardt as generous readers and interlocutors at various stages of the process. At Rutgers, I was fortunate to be a member of the Center for Cultural Analysis's research seminar Object and Environment where I had thought-provoking conversations with Anita Bakshi, Max Hantel, Colin Jager, Jorge Marcone, Margaret Ronda, Eric Sarmiento, Sean Tanner, Laura Weigert, and Darryl Wilkinson.

I presented materials from this book in universities, activist spaces, and cultural institutions and received invaluable feedback and helpful critiques. These include the University of Bologna, the University of Bern, the University of Cambridge, the Graduate Institute in Geneva, La Sapienza University, Saint Gallen University, Esc Atelier, and Lucha Y Siesta in Rome.

I am grateful to my colleagues Donatella Della Ratta, Antonio López, Peter Sarram, and Anthony Stagliano for welcoming me into the department of Communication and Media at John Cabot University in Rome. The University of Lausanne and the Institute of Geography and Sustainability provided me with the time and resources to write this book. A special thanks there goes to Valérie Boisvert, Irene Becci, Joëlle Salomon Cavin, Julia Steinberger, Mathis Stock, and Gretchen Walters. Thank you also to the students of the graduate courses Ideas and Politics of Nature and Environmental Justice: Transnational Perspectives for their critical curiosity and to my teaching and research assistants, particularly Alberto Manconi.

Thanks to my friends in New York City, Boston, and other parts of the United States for many convivial moments, sustaining relationships, and stimulating intellectual and political challenges throughout the years: Carolina Alonso Berejano, Jack Bratich, Katherine Behar, Beka Economopoulos, James Hay, Malav Kanuga, Jason Jones, Lize Mogel, Joanne Morreale, and Stina Soderling. The activist hubs 16 Beaver and Not An Alternative enriched my life and thinking through countless seminars and public events. Loving thoughts to my friends in Rome and particularly Marina Montanelli, Francesco Raparelli, Valeria Ribeiro, Barbara Bonomi Romagnoli, Laura Ronchetti, Agnese Trocchi, Marina Turi, and all those who I have not remembered to acknowledge, and I hope will forgive me for that. I am grateful to my mother Franca for her love and attachment to life; to my brothers Alessandro, Massimo, and Roberto; to Teo for taking care of my mother; to Sardinia, the island of rocks, wind, sand, and water that shaped me with its asperities and silences. Special thanks to Marco Deseriis and Tito Tola Deseriis: *con amore*; to Tina Aquili and Giovanni Colombo for their support; to Jason and Timer, furry companions, for demonstrating the importance of eating, resting, and sleeping well.

Laura Portwood-Stacer and Meghan Drury provided valuable editorial feedback throughout various stages of the manuscript's development, particularly the book proposal and the first draft. At Fordham University Press, I especially appreciate Tom Lay's enthusiasm and steadfast support as I completed the book. Bruce Clarke and Henry Sussman have warmly welcomed *Resurgent Commons* into the wonderful book series Meaning Systems, and I'm honored to be part of it. My heartfelt thanks go to the anonymous reviewers for their time, labor, and insightful comments on the manuscript. I am grateful to Kem Crimmins for guiding this book through production, and to Lis Pearson for excellent copyediting. I gratefully acknowledge photographer Giordano Pennisi and the Maria Baccante Documentation Center in Rome for granting permission to use the cover images.

Small portions of this book appeared in earlier publications. Chapter 3 includes significantly revised sections of my article "Species, Nature and the Politics of the Common: From Virno to Simondon," published in *South Atlantic Quarterly* 116, no. 2 (2017). An earlier version of Chapter 5 appeared in *Environmental Humanities* 11, no. 1 (2019) with the title "The Archive and the Lake: Labor, Toxicity, and the Making of Cosmopolitical Commons in Rome, Italy." I am grateful to the publishers and editors of these publications.

NOTES

INTRODUCTION: RESURGENCES

1. Stephanie LeMenager (2021) also mentions Robinson's novel in a chapter centering the commons in the environmental humanities. "The work of artists and literary authors—writes LeMenager—figures here as a seedbed for politics" (12). She notes that by setting the novel in a submerged city Robinson shifts from land to water as "a newer ontological ground, for lives unmoored in flooding" (22).

2. Rome is the city I moved to from the Mediterranean island of Sardinia to attend university, and the one I returned to after a decade in the United States. For me, it is a place of dense socio-ecological relations and activist kinships.

3. The Anthropocene Working Group (AWG), chaired by the geologist Jan Zalasiewicz, was established in 2009 by the Subcommittee on Quaternary Stratigraphy within the International Union of Geological Sciences (IUGS). After heated debates, including controversies concerning the composition of the panel, made up mainly by Western male geoscientists, the AWG identified the start of the Anthropocene, the so-called Golden Spike, in the mid-twentieth century, when nuclear bomb tests left radioactive fallout across the planet, and proposed to officially declare the Anthropocene as a new epoch in the earth's timeline. In March 2024, the IUGS rejected the proposal. The controversy, within and outside the geoscience community, however, is far from over.

4. On the difference that feminism makes in the environmental humanities see Jennifer Hamilton and Astrida Neimanis (2018).

5. Grounding my analysis of more-than-human urban commons in complex histories beyond the European space (see Chapters 2 and 6), I join urban studies scholars Tessa Eidelman and Sara Safransky in arguing for the need to "grapple more deeply with questions of race, gender and (de)colonization" (2020, 799). As they eloquently articulate, this is relevant to make sense of ongoing enclosure dynamics and finding ways to resist them.

6. Autonomist Marxism, and the tradition of *operaismo* (workerism) that preceded

it, is a peculiar form of theorizing *within social movements* that emerged in Italy during a long season of struggle that began in the 1960s. For an introduction to the workerist project and autonomist movements in English see Sylvere Lotringer and Christian Marazzi (1980), Paolo Virno and Michael Hardt (1996), Steve Wright (2002), and Gigi Roggero (2023).

7. I borrow this formulation from Philippe Pignarre and Isabelle Stengers who raise the question of inheriting from Marx in the book *Capitalist Sorcery* (2011). "The crucial point," they write "is not to reach agreement on what Marx wrote, but to prolong the question that he *created*, that of this capitalism whose hold is a matter of combating" (12). Moreover, inheriting from Marx means to prolong his questions in the presence of those who have collided with the Marxists: "those groups in struggle, who, like the feminists, refused the order of priorities proposed in the name of class struggle; like the radical ecologists, who have had to struggle against the assimilation of nature to a set of resources to be valorized; like the peasants, who have had enough of the taste of productivism; like the indigenous people who have to deal with the unanimous judgment that identified their practice with simple superstitions, etcetera" (11).

8. Drawing on Federici, the political theorist Verónica Gago has explored the concept of "fascistization of social reproduction" to describe the intensification of value extraction under the presidency of Javier Milei in Argentina and the reelection of Donald Trump in the United States. Emphasizing the central role of anti-feminist and anti-trans policies in these far-right administrations, Gago suggests that their rise is connected to a reactive response to the recent repoliticization of social reproduction and care by transfeminist movements over the past decade. See Gago (2025).

9. My deepest gratitude goes to the Kanaka Maoli scholar J. Kēhaulani Kauanui for sharing her unpublished work on the commons with me.

10. On the different positionalities of settlers and Indigenous people with respect to the commons in North America see also Craig Fortier (2017). Showing that radical left movements' struggles for the commons have often erased Indigenous claims to land, Fortier argues that commons are multiple and at times incommensurable.

1. READING ACROSS COMMONS ARCHIVES

1. Isabelle Stengers insists on the importance of slowing down in order to think and act differently in the face of vast socio-ecological problems. See Stengers (2005).

2. Jordy Rosenberg (2014) has argued that new materialism and the ontological turn frequently "drive toward the occlusion of the dynamics of social relation" through the "de-suturing of objects from the social world, an unloosing of the socius from historical time" (n.p.). Other scholars have critiqued the turn toward matter and materiality for the failure to engage with racial dynamics of objectification (Jackson 2015) and the oversight of a new rationality of government and mode of accumulation that thrives on ever-changing natures (Pellizzoni 2015; 2025). These arguments are not without merits.

3. For an account of the juridical proximity of divine things and public things in Roman law see Yan Thomas (2002). See also Roberto Esposito (2015).

4. In *Spiritual Franciscans: From Protest to Persecution in the Century After Saint Francis*, historian David Burr provides a detailed account of Peter Olivi's work and his connection to Joachim of Fiore, a key figure in the apocalyptic tradition concerned with the end of the world. Burr also chronicles the Catholic Church's campaign against the Spirituals and their influence on heretic movements. See Burr (2001).

5. Giorgio Agamben's emphasis on the radical Franciscans' renunciation of property and the right to possess departs from the influential study by Paolo Grossi (1972) who traces the origins of subjective rights to Franciscan texts.

6. Agamben elaborates on the theme of use in *The Use of Bodies* (2015), the book that brings to an end the *Homo Sacer* series.

7. I return to Marx in Chapter 2 to consider his conception of labor in relation to the natural world.

8. In *The States of the Earth*, Mohamed Amed Meziane observes that primitive accumulation was sustained by the disenchantment of the soils and the subsoils. The state confiscation of church lands in sixteenth-century England led to the expansion of mining and, eventually, "to a dynamic of primitive accumulation of fossil capital . . . which involved a new way of understanding the subterranean. . . . As industrial and fossil capitalism developed, representations of a subterranean world peopled with a multitude of extra-human forces were discredited as the superstitious beliefs of the lower classes" (Meziane 2024, 163–64). Part of the long process of secularization, this dynamic would be extended to the European colonies in the nineteenth century when fossil empires relied on a racialized labor force for their extractive operations. Secularization, Meziane argues, is not the decline of religion. Rather, it is "the fulfilment of religion in this world by transforming nature through industry" (ibid., xiv). I am grateful to Tom Lay for drawing my attention to Meziane's brilliant work.

9. Carlo Ginzburg's study was first published in Italian in 1976 and translated into English in 1980. This highly acclaimed and much debated example of microhistory used the Inquisition's trial records and other documents to trace the cosmology of the subaltern.

10. Commenting on Caroline Bynum's formulation of "holy matter," the medievalist and environmental humanities scholar Jeffrey Jerome Cohen points out that "In the Middle Ages materiality exists to reveal something about God, certainly, but it also reveals something about itself, something that cannot be wholly subsumed into allegory" (Cohen 2015, 13).

11. Maria Mies, in particular, has had a significant influence on Silvia Federici's work. For Mies, the primitive accumulation of capital on a global scale in early modern times depended on the persecution of women as witches and processes of enslavement and extraction in the European colonies. For Mies "we cannot understand the modern developments, including our present problems, unless we include all those who were 'defined into nature' by the modern capitalist patriarchs: Mother Earth, Women and Colonies" (Mies 1989, 75).

2. TRANSATLANTIC ECOLOGIES OF DISPOSSESSION

1. For an excellent discussion of how modern political ideas depend on collective relations to the material world see Pierre Charbonnier (2021).

2. Michael Best's edition of Gervase Markham's compendium provides a useful introduction to this work, its intellectual tradition and sources. See Markham (1986).

3. A classic ecofeminist critique of gendered nature and masculine science in Francis Bacon is Carolyn Merchant's *The Death of Nature. Women, Ecology and the Scientific Revolution* (1980). See also Evelyn Fox Keller's chapter "Baconian Science: The Arts of Mastery and Obedience," in *Reflections on Gender and Science* (1985).

4. Brenna Bhandar observes that while in William Petty's times biological racism had yet to emerge, notions of racial difference and European superiority were tied to specific methods for making land productive: "my argument is that racial subjects and modern property laws are produced through one another in the colonial context. . . . The types of use and possession of land that justified ownership were determined by an ideology of improvement" (2018, 8). Bhandar contends that this ideology has been adapted in a range of colonial contexts, from British Columbia in the nineteenth century to Palestine in the twentieth century when Zionist settlers sought to disqualify Arab methods of cultivation and use the modernization of agriculture as basis for establishing their rights on Palestinian land.

5. Barbara Arneil maintains that John Locke's chapter on property in *Two Treatises of Government*, "was written to justify the seventeenth-century dispossession of the aboriginal peoples of their land, through a vigorous defence of England's 'superior' claims to proprietorship" (1996, 2). She also discusses Locke's racial hierarchy between Amerindians, seen as primitive, "pre-Christian man" and Black Africans, considered as less than human. This distinction allowed him to make profits from the slave trade through the financial investment in the Royal Africa Company.

6. The political theorist Onur Ulas Ince develops a compelling reading of money in Locke's liberal justification of English colonialism. He argues that Locke predicated labor and improvement on monetization and characterized the absence of money in America as a clear sign that the New World remained a common open to appropriation. See Ince (2018).

7. The debate on the metabolic rift has been recently revived by the popularity of Kohei Saito's important reading of Marx's scientific notebooks and letters from the 1870s in which, among other things, he discusses the use of fertilizers in agriculture. Saito argues that in this period Marx revised his critique of political economy by centering the concern for the biophysical limits to capitalist growth. While I admire the erudition of this analysis, ultimately, I remain unconvinced by the characterization of Marx as a philosopher of ecological limits. Moreover, establishing the truth about Marx's interpretation, beyond the "misunderstandings" that Saito in part attributes to Friedrich Engels's editorial work, does not change the productivism that has long prevailed in Marxism. See Saito (2017).

8. For a critical reading of Marx's materialist understanding of the relationship between society and nature see also Luigi Pellizzoni (2025).

3. ENGAGING POTENTIALS AND LIMITS OF THE MARXIST COMMON

1. Consider, for example, *The Economist*'s cover story on the Anthropocene in 2011, and the Canadian artist Edward Burtynsky's "The Anthropocene Project," comprising fine arts photography, a documentary film, and several VR short films that travel galleries and museums around the world.

2. The seeds of the debate on cognitive capitalism were planted in the publication of *Futur Antérieur*, a journal cofounded by Antonio Negri in 1989, after the fall of the Berlin Wall. For a decade the journal brought together French and Italian militant intellectuals researching the metamorphosis of labor under capitalism and the forms of struggles arising in this context. The group formed around *Futur Antérieur* included the economist Yann Moulier Boutang, the author of the book *Cognitive Capitalism* (2012) who defined it as "a mode of accumulation in which the object of accumulation consists mainly of knowledge" (57). As cognitive capitalism reorganizes the processing of raw materials and manufacturing, financialization becomes "the expression of this remodeling, of this reformatting, of material production" (48). For a concise and effective introduction to cognitive capitalism and its critics see David Harvie and Ben Trott (2021).

3. For Antonio Negri's first-hand account of the imprisonment, the trial, and the decision to leave Italy see Negri (2010).

4. In *The Condition of Postmodernity*, David Harvey (1991) discusses the crisis of the Fordist model in the mid-1960s. He argues that "these difficulties could be best captured by one word: rigidity" (142). He describes the combination of rigid long-term and mass-scale mass production systems, rigid labor markets and contracts, and the seemingly fixed configuration of political power. For Harvey and many other scholars, the capitalist response to the crisis was the transition to a new regime of flexible accumulation resting on "flexibility with respect to labor processes, labor markets, products and patterns of consumption" and combined with new ways of providing financial services and the intensification of technological and organizational innovation (147).

5. The interview, conducted in Paris in March 2022, is available online. See https://www.youtube.com/watch?v=qef6M2mYoFA.

6. For a different critical take on life, autonomy, and antagonism in autonomist Marxism see Elizabeth Povinelli (2016). This essay draws on the author's relation with long term Indigenous collaborators from the coastal lands in the Top End of the Northern Territory, Australia.

7. On the relationship between race and the animal and the exclusion of racialized and gendered bodies from the category of the human see the brilliant contributions of Bénédicte Boisseron (2018) and Zakiyya Iman Jackson (2020).

8. For an expanded analysis of Paolo Virno's work in relation to Gilbert Simondon see Miriam Tola (2017).

9. Simondon's first publication, *Du Mode d'existance des objects techniques* appeared in France in 1958. It was only in 1989, with the posthumous release of *L'individuation psychique and collective*, that his work began to be widely read. His writings on individuation have been published as a whole in *L'individuation a la lumière des notions de forme et d'information* (Simondon 2005).

10. For the purposes of this book Muriel Combes's (2013) argument about Simondon's non-anthropocentric anthropology is particularly relevant. See also the feminist readings of Simondon by Hasana Sharp (2011) and Elizabeth Grosz (2012).

11. Initially Michel Foucault used the concepts of biopower and biopolitics to describe the ways in which political authorities invest in the bodily vitality of populations through a series of calculations, interventions, and regulatory controls. In *Security, Territory, Population* (2007), however, he specifies that the biopolitical fostering of life was the crucial preoccupation of liberal governmentality, a form of power that emerged in Europe in the eighteenth century and which anticipated many features of contemporary governance.

12. Alberto Toscano notes that Foucault presents biopolitics as an "anti-universal" concept. According to Toscano, Foucault's decision to speak of "governmental practices" is a way to put aside universals such as the state, society, and the sovereign. Rather than assuming the existence of such universals, Foucault's genealogical approach describes new governmental practices as a field shaped by heterogeneous forces and technologies. See Toscano (2007a).

13. Particularly in current neoliberal authoritarian contexts, the feminist politics of *denaturalization* remains crucial for doing away with gendered and racialized ideologies of nature as a normative realm. However it present limits for grasping the dynamics of environmental injustices and exploring alternative socio-ecological and multispecies relations.

14. Emanuele Leonardi's book *Lavoro, Natura, Valore* (2017) provides an important contribution to the debate on the valorization of nature. Leonardi argues that the labor-nature-value nexus has shifted from a configuration in which external nature is appropriated and transformed by labor to generate value to a new one in which capital also finds ways to internalize nature and generating profit directly from it. The creation of carbon trading markets is an example of this shift.

15. In posing these questions, I adapt Philippe Pignarre and Isabelle Stengers's reflection on how to inherit from Marx's critique of capitalism without reproducing the Marxist grand narrative of history. See Pignarre and Stengers (2011).

4. TRANSFEMINIST COMMONS: INHABITING THE EARTH OTHERWISE

1. Reflections on feminist and queer commons have flourished in activist and academic circles in various contexts. Silvia Federici's seminal work has resonated deeply with activist milieus in Europe and the Americas (Federici 2004; 2018). In Latin America, Raquel Gutiérrez Aguilar has developed a feminist approach to the commons as a praxis for defending life, both human and non-human (Aguilar et al. 2016). J. K. Gibson-

Graham (2016) has linked commoning to postcapitalist community economies that challenge the prevalent understanding of capitalism as ubiquitous and ever-powerful. The field of feminist political ecology has taken up these topics and explored commoning in relation to uneven embodied experiences of environments (Clement et al. 2019). Queer studies scholarship has challenged normative assumptions within and about the commons. Lauren Berlant (2016) reframes it as a "pedagogy of unlearning" with respect to normative infrastructures. José Esteban Muñoz (2013) has gestured toward queer "brown" commons in which Brown people, places, animals, and minerals suffer together but also cultivate the ability to flourish under duress. See also the 2018 special issue on "The Queer Commons" in *GLQ: A Journal of Lesbian and Gay Studies*.

2. It is worth recalling that the Council of Europe recommends having at least one accommodation in a shelter per 10,000 people as the minimum standard for combating violence against women in European countries. In Rome, a city that would need about four hundred accommodations, only thirty beds were available in 2020 and fourteen of them were at Lucha y Siesta. This means that over the years the city's administration has benefited from the feminist shelter's social services, which activists say have saved the city over six million euros.

3. The artists supporting Lucha y Siesta through a call launched by Rita Petruccioli have created playful visual interpretations of the figure of the *luchadoras*, the women wrestlers of the Mexican sporting spectacle of *lucha libre*.

4. The term transfeminism as I use it in this book has been circulating since the early 2000s in Southern Europe and Latin America. It refers to capacious activist and intellectual frameworks informed by feminist, queer, and trans critiques of patriarchal violence, heteronormativity, and gender binarism. Importantly, this approach is attentive to the uneven transnational dimensions of patriarchal and capitalist formations. According to the Mexican scholar Sayak Valencia, transfeminism entails troubling the political subject of feminism to address the necropolitical effects of capitalist and state violence in localities differently marked by histories of dispossession and social exclusion (Valencia 2018). For a compelling account of transfeminism in Southern Europe with a focus on Italy see Elia Arfini (2022). See also Michela Baldo (2019).

5. My own thinking about care during the pandemic developed through exchanges with many collaborators and friends, particularly Maddalena Fragnito with whom I coedited the volume *Ecologie della cura. Prospettive transfemministe* (2021).

6. The term feminist constellation is a reference to the book *8M Constelación feminista*, published in 2018 and translated into Italian and German. It included short chapters by authors based in Argentina, Italy, Mexico, United States, and Uruguay. See Gago et al. (2018).

7. Brunella Casalini argues that while the feminist ethics of care demonstrates the everyday complexities of care relations, feminist theories of social reproduction enable us to grasp the dynamics of care as work in global capitalism. See Casalini (2018). Writing with Maddalena Fragnito, I suggest that "it would be short-sighted to dismiss the ethics of care as carrying sentimental residues that would be a distraction in the process of advancing an anti-capitalist political project. It would be equally counter-

productive to evoke the centrality of care without focusing on the ways in which it is distributed, privatized, unpaid and denied in the uneven geographies of patriarchal and racial capitalism. Rather than ignoring the tension, it is therefore worth asking what these conceptual grids have to offer each other" (Fragnito and Tola 2021, 22).

8. Other critics have addressed the binaries of labor and care, the human and the nonhuman and heteronormative sex/gender in social reproduction literature. For a "sideways" approach to social reproduction see Andrucki et al. (2017). See also Malatino (2020).

9. Discussions of ecofeminism in academia have long been characterized by a sense of unease about the association between women and nature. In "Ecofeminism Revisited" (2011), Greta Gaard provides a stimulating account of these debates and the contributions of ecofeminism from the 1980s onward. Emphasizing the ecofeminist insights on the connections among racism, sexism, and the appropriation of nature, she proposes to retrieve ecofeminism as foundational for much of current feminist scholarship of socio-natural entanglements. On ecofeminism see also the important interventions by Catriona Sandilands (1999), Sherilyn MacGregor (2006), and Charis Thompson and MacGregor (2017).

10. My disagreement with some ecofeminist characterizations of the earth as a homeostatic system is informed in part by Ilya Prigogine and Isabelle Stengers's *Order out of Chaos: Man's New Dialogue with Nature* (2018 [1984]). Here they provide insights into nature and matter as unstable, characterized by the sudden and irreversible emergence of organized structures out of highly turbulent situations. According to Prigogine and Stengers, the rearticulation of time and nature allows us to shift the view of molecular processes that have enabled the emergence and variation of life forms on earth "toward the multiple, the temporal, and the complex" (292). Yet, I am also mindful of Stengers's political convergence with ecofeminist approaches that connect the relentless capitalist plunder of the earth to gendered forms of dispossession. For an analysis of Stengers's feminist politics of the earth see Tola (2016).

11. Sara Nelson's lucid analysis of the ecosystem service economy makes an explicit reference to Federici's work. Nelson revises autonomist Marxism to investigate how in ecosystem services the activities of both social and ecological reproduction become direct sources of value. See Nelson (2014).

12. *Life Beyond the Pandemics* was first circulated through social media and activist media outlets in Italian and later published in English in *Interface: A Journal For and About Social Movements*. The citations refer to the English version.

13. B-side Pride, "Non saremo 'congiunti', ma unite nella lotta," https://smaschieramenti.noblogs.org/post/2020/04/27/non-saremo-congiunti-ma-unite-nella-lotta/.

14. For an excellent English translation of the debate on body-territory as a feminist decolonial praxis and method see Cruz et al. (2022).

15. The *Dichiarazione di Autogoverno* (in Italian) is available on Lucha Y Siesta website at https://luchaysiesta.org/la-dichiarazione-di-autogoverno/.

16. The Black Mediterranean Collective (2021) provides important insights on how colonialism and racial capitalism shape the dynamics of exclusion and abandonment

in Europe. On issues of race, gender, and citizenship in contemporary Italy see also Angelica Pesarini (2023).

17. On prefigurative politics see David Graeber (2013). For an overview of the concept see Lara Monticelli (2022).

18. Malcom Ferdinand provides a powerful critique of colonial habitation from the vantage of the Caribbean. He details how racism and colonialism have been functioning as "a way of inhabiting the Earth that includes an engineering of its environmental, social, and political landscapes" (2021, 178).

19. My approach to transfeminist commons intersects Michele Lancione's proposition for a liberatory politics of home. Lancione writes, "Going beyond the violence of mainstream inhabitation is a creative act: it is about providing the ground upon which collectives can self-define, self-liberate, and otherwise care about their inhabitation of the planet" (2024, 214).

5. COSMOPOLITICAL COMMONING IN A CITY OF RUINS

1. On cosmopolitics see Isabelle Stengers (2005). I borrow the notion of "becoming with" from Donna Haraway (2016).

2. For an account of bodies' porosity and the relationship between bodies and environments see Stacy Alaimo's concept of transcorporeality (2010).

3. Note the diagnosis of hysteria here. As feminist historians have shown, in the nineteenth century, hysteria became a keyword in Western medical lexicon to make sense of shifts in gender and race relations in contexts of rapid industrialization and urbanization. See Laura Briggs (2000).

4. The medical anthropologist Niso Tommolillo (2020), a cofounder of the Ex-SNIA archive, combines archival research and narratives techniques to convey the stories of chemical harm taking place in Rome's rayon factory.

5. Matilde Fracassi, interview with the author, May 3, 2017.

6. Writing about socio-environmental struggles over toxic waste in Naples, Italy, Marco Armiero and Salvatore Paolo De Rosa (2016) argue that the sense of smell can trigger political activism. My focus here is the role of bodily sensations in nourishing political activism across generations.

7. Alessandra Conte, interview with the author, May 19, 2017.

8. The medical data preserved at the Ex-SNIA archive are the product of managerial control of workers' health. But the constitution of the archive privileges critical viewpoints on factory work and its role in structuring neighborhood life. In grappling with the ethical implications of researching medical information regarding people who worked in the factory, I followed the activist practice of identifying them by initials only. Further, this chapter and all my writings about the EX-SNIA have been shared with the local activist community.

9. SNIA Medical Register, accessed by the author on June 14, 2017 at the historical archive Centro di documentazione Maria Baccante—Archivio Storico Viscosa.

10. Forum Parco delle Energie, "Storia dell'area," lagoexsnia.files.wordpress.com /2014/01/allegato_8.pdf (accessed on November 20, 2023).

11. Alessandra Conte, interview with the author, May 19, 2017.

12. Alessandra Conte, interview with the author, May 19, 2017.

13. SNIA Workers Files, accessed by the author on May 10, 2017 at the historical archive Centro di documentazione Maria Baccante—Archivio Storico Viscosa.

14. Matilde Fracassi, interview with the author, May 3, 2017.

15. The work of Franco "Bifo" Berardi offers a useful contribution for exploring contemporary forms of toxic labor and psychic distress in post-Fordist contexts of uneven precarity. See, for example, Berardi (2012).

16. Flavia Sicuriello, interview with the author, May 10, 2017.

17. Alessandra Conte, interview with the author, May 19, 2017.

18. In using the terms ecological repair and reparation I am taking a cue from Maria Puig de la Bellacasa and Dimitris Papadopoulos's video project on these topics and the corresponding book (Papadopoulos et al. 2023). Ecological reparation has the twofold meaning of both repairing and reclaiming damaged ecologies and transforming damaging socio-ecological relations.

19. April 25, the day that marks Italy's liberation from fascism, is usually celebrated with a parade in the Prenestino-Labicano neighborhood. In 2016 this parade culminated in the lake's "liberation," highlighting how anti-fascism has always been central to the Ex-SNIA struggle.

20. The legal definition of "natural monument" has been adopted internationally for protecting natural/cultural landscape features of outstanding value for their rarity, aesthetic qualities, or cultural significance. These may include submarine caverns, trees, and geological formations. For a definition of "natural monuments" by the European Environmental Agency see https://www.eea.europa.eu/help/glossary/eea-glossary /natural-monument. For a discussion of the Bullicante Lake from a legal studies perspective, see Petersmann (2024).

21. Matilde Fracassi, interview with the author, May 3, 2017.

22. See Chapter 2 for my analysis of Marx's notion of metabolism.

23. Alessandra Conte, interview with the author, May 19, 2017.

24. Combining Alfred North Whitehead's speculative inquiry and field philosophy, Valeria Cirillo's PhD dissertation, successfully defended in 2024, provides rich insights on the more-than-human attachments in two postindustrial humid zones: the Ex-SNIA in Rome and the Marais Wiels in Brussels. An activist-scholar, Cirillo also contributed to their encounter. See Cirillo (2024). On the notion of attachment, see also Stephen Muecke (2017).

25. The video for *Il lago che combatte* is available here: https://www.youtube.com /watch?v=Dcb_Thrq2P8. For a discussion of this and other Ex-SNIA's media artifacts from a ecomedia perspective, see Tola (2024).

6. CROSSING THE STORM: ENCOUNTERING DECOLONIAL PERSPECTIVES

1. On gender and coloniality see the groundbreaking work by Maria Lugones (2007) and its problematization by Catherine Walsh (2016).

2. Since this gathering was not public and I did not ask for permission to share the content of the dialogue between the Zapatista and the local activists, here I am only offering some general observations without entering into the details of the conversation. The stories that were shared there are not mine to disclose.

3. In reflecting on the overuses of the "decolonial" in academia, I am inspired by remarkable studies by Eve Tuck and K. Wayne Yang (2012) and Max Liboiron (2021).

4. Nandita Rani Sharma has since developed a critique of Indigenous claims to sovereignty and offers a decolonized commons as a formation alternative to national self-determination. See Sharma (2020).

5. See Chapter 2 for a historical discussion of the deployment of the commons in settler colonial ecologies of dispossession.

6. At the most basic level political ontology describes what constitutes a political collective and the power relations that traverse it. It consists of "the power laden negotiations involved in bringing into being the entities that make up a particular world" (Blaser 2013, 11). Ontologies precede mundane practices but are constantly reshaped by them.

7. The edited collection *The Anomie of the Earth* (Luisetti et al. 2015) makes an important step toward bringing into conversation Latin American decolonial thought and Italian autonomist theories of autonomy and the commons. As stated by the volume's editors in the introduction, the goal is fostering a "conceptual dialogue and political alliance between contemporary movements of de-Westernization and the resistance against capitalist labor and biopower coming from workerism and post-autonomia: in both instances, a line of flight from the central institutions and commitments of Western modernity is calling into question century-old habits of thoughts and political action, proposing concepts and practices that bypass the lexicon of political modernity and academic cosmopolitanism" (3). This chapter extends this effort while also highlighting tensions between these intellectual and political projects.

8. For an assessment of the Women's Revolutionary Law twenty years after the uprising of 1994, see Sylvia Marcos (2014).

9. Subcomandante Marcos's speech is available at http://inthesetimes.com/issue/25/10/marcos2510.html.

10. The third communiqué focuses on the intergenerational dimension of the Zapatista's project of crossing the storm. It is available at this link: https://enlacezapatista.ezln.org.mx/2023/11/02/tercera-parte-deni/ (accessed on June 11, 2024).

11. The English translation of the communiqué is available at this link: https://enlacezapatista.ezln.org.mx/2023/12/22/twentieth-and-last-part-the-common-and-non-property/ (accessed on November 11, 2024).

EPILOGUE: ON INSTITUTING

1. It is worth noting that in *New York 2140* it is the United States that leads change on climate policy thus somehow supporting the idea of North American geopolitical exceptionalism. In contrast, a rich body of decolonial scholarship and activism foregrounds how Southern "submerged perspectives" (Gómez-Barris 2017) emerging from "extractive zones" in Latin America and Africa have been instrumental for creating cartographies of planetary extractivism and weave alternative social ecologies.

2. On the relationship between authoritarianism and neoliberalism, see Quinn Slobodian (2025) and Catherine Rottenberg (2025). On microfascism and its deadly gendered manifestation see Jack Z. Bratich (2022).

3. Shourideh Molavi (2024) offers a poignant spatial analysis of how Israel's settler colonialism has been operating through Palestinian environments and Palestinian landscapes of anti-colonial resistance.

4. Sara Farris (2024) has read Giorgia Meloni's rise to power through what she terms femonationalism, that is the instrumentalization of feminist issues by nationalism.

5. For a compelling debate on the relationship between states and social movements at the time of heightened climate injustices, see the recent forum "Climate, State, and Utopia" published in the *Boston Review* (Spring 2024) with an opening contribution by Olúfẹ́mi O. Táíwò.

6. My argument in part intersects with the proposition for a liberatory politics of the home advanced by Michele Lancione (2024). For Lancione, this consists of three movements. The first is deinstitution, that is, the fight against the industries providing services for the homeless and producing knowledge about them. The second is reinstitution through the reappropriation of the capacity of caring for inhabitation and the development of a collective praxis of radical care that addresses the violence of mainstream politics of the home. Reinstitution is enacted by and with dwellers who refuse to be categorized simply as people in need of top-down policy interventions. Finally, institution is the creative moment of affirmation. It is about "bypassing state inertia and prefiguratively enacting a liberatory politics of the home in the here and now" (18) through grassroots organizing, squatting, and the proliferation of ways of homing.

7. Francesco Raparelli's view of institutions in the book *Singolarità e istituzioni* (2021) combines philosophical interpretation and insights from social movements and grassroots trade unions. It diverges from Giorgio Agamben's perspective on destituent power which, in refusing the state, also refuses institutions thus conflating them with sovereign transcendence. Instead, it is closer to Antonio Negri's constituent power because it conceives institutions as produced through the insurgencies of living labor. It is worth noting that another Italian philosopher, Roberto Esposito, has identified a third space between Agamben and Negri. Esposito (2022) argues that those who present institutions as established solely around state power obscure the intimate relationship between life and institutions. His genealogy of the ancient Roman concept of

vitam instituere seeks to shed light on this connection. Politically, however, Esposito's proposition is perplexing. After describing the decline of state sovereignties, he seems to embrace the rise of new political institutions (primarily NGOs) as the expression of global civil society.

WORKS CITED

Acosta, Alberto. 2010. *El Buen Vivir en el camino del post-desarollo. Una lectura desde la Constitucion de Montecristi*. Frederich Ebert Stiftung, Policy Papers 9. Quito: Ecuador FES-ILDIS.

Agamben, Giorgio. 2013. *The Highest Poverty: Monastic Rules and Form-of-Life*. Stanford University Press.

Agamben, Giorgio. 2015. *The Use of Bodies*. Stanford University Press.

Agard-Jones, Vanessa. 2014. "Spray." Somatosphere. Retrieved from http://somatosphere.net/2014/05/spray.html.

Aguilar, Raquel Gutiérrez, Lucia Linsalata, and Mina Lorena Navarro Trujillo. 2016. "Producing the Common and Reproducing Life: Keys Towards Rethinking the Political." In *Social Sciences for Another Politics*, edited by Ana Cecilia Dinerstein, 79–92. Palgrave Macmillan.

Alaimo, Stacy. 2010. *Bodily Natures: Science, Environment, and the Material Self*. Indiana University Press.

Allen, Catherine. 2002. *The Hold Life Has: Coca and Cultural Identity in an Andean Community*. Smithsonian Institution Press.

Andolina, Robert, Nina Laurie, and Sarah A. Radcliffe. 2009. *Indigenous Development in the Andes: Culture, Power, and Transnationalism*. Duke University Press.

Andrucki, Max, Caitlin Henry, Will McKeithen, and Sarah Stinard-Kiel. 2017. "Intro: Beyond Binaries and Boundaries in 'Social Reproduction.'" *Society & Space*. Retrieved from https://www.societyandspace.org/forums/beyond-binaries-and-boundaries-in-social-reproduction.

Arboleda, Martín. 2020. *Planetary Mine: Territories of Extraction Under Late Capitalism*. Verso Books.

Arfini, Elia. 2022. "Italian Queer Transfeminism Towards a Gender Strike." In *LGBTQ+ Intimacies in Southern Europe: Citizenship, Care and Choice*, edited by Ana Cristina Santos, 233–51. Palgrave Macmillan.

Armstrong, Regis, J. Wayne Hellman, and William J. Short, eds. 2000. *Francis of Assisi: Early Documents, Volume II: The Founder*. New City Press.

Armiero, Marco, and Salvatore Paolo De Rosa. 2016. "Political Effluvia: Smells, Revelations, and the Politicization of Daily Experience in Naples, Italy." In *Methodological Challenges in Nature-Culture and Environmental History Research*, edited by Jocelyn Thorpe, Stephanie Rutherford, and L. Anders Sandberg, 173–86. Routledge.

Armiero, Marco, and Massimo De Angelis. 2017. "Anthropocene: Victims, Narrators, and Revolutionaries." *South Atlantic Quarterly* 116 (2): 345–62.

Armiero, Marco, and Leandro Sgueglia. 2019. "Wasted Spaces, Resisting People: The Politics of Waste in Naples, Italy." *Revista Tempo e Argumento* 11 (26): 135–56.

Arneil, Barbara. 1996. *John Locke and America: The Defence of English Colonialism*. Clarendon Press.

Asara, Viviana. 2020. *Democrazia senza crescita. L'ecologia politica del movimento degli Indignados*. Aracne.

Augustine. 1998. *The City of God Against the Pagans*. Cambridge University Press.

Bacon, Francis. 2002. *The Major Works*. Oxford University Press.

Ballard, James Graham. 1965. *The Drowned World and the Wind from Nowhere*. Doubleday.

Baldo, Michela. 2019. "Translating Spanish Transfeminist Activism into Italian: Performativity, DIY, and Affective Contaminations." *Gender / Sexuality / Italy* 6 (1): 66–84.

Barca, Stefania. 2012. "Bread and Poison: Stories of Labor Environmentalism in Italy 1968–1998." In *Dangerous Trade: Histories of Industrial Hazard Across a Globalizing World*, edited by Christopher Sellers and Joseph Melling, 126–39. Routledge.

Barca, Stefania. 2014. "Telling the Right Story: Environmental Violence and Liberation Narratives." *Environment and History* 20 (4): 535–46.

Barca, Stefania. 2020. *Forces of Reproduction*. Cambridge University Press.

Bartlett, Robert. 2008. *The Natural and the Supernatural in the Middle Ages*. Cambridge University Press.

Battacharya, Tithi. 2017. *Social Reproduction Theory: Remapping Class, Recentering Oppression*. Pluto Press.

Battisti, Corrado, Giuseppe Dodaro, and Giuliano Fanelli. 2017. "Paradoxical Environmental Conservation: Failure of an Unplanned Urban Development as a Driver of Passive Ecological Restoration." *Environmental Development* 24: 179–86.

Battistoni, Alyssa. 2017. "Bringing in the Work of Nature: From Natural Capital to Hybrid Labor." *Political Theory* 45 (1): 497–517.

Bennett, Jane. 2010. *Vibrant Matter. A Political Ecology of Things*. Duke University Press.

Bennett, Jane. 2013. "The Elements." *Postmedieval* 4 (1): 105–11.

Berardi, Franco B. 2012. "Exhaustion/Depression." In *Depletion Design: A Glossary of Network Ecologies*, edited by Carolin Wiedemann and Soenke Zehle, 79–84. Institute of Network Cultures.

Berardi, Franco B. 2013. "Accelerationism Questioned from the Point of View of the Body." *e-flux Journal*, 46. Retrieved from https://www.e-flux.com/journal/46/60080/accelerationism-questioned-from-the-point-of-view-of-the-body/.

Berlant, Lauren. 2016. "The Commons: Infrastructures for Troubling Times." *Environment and Planning D: Society and Space* 34 (3): 393–419.

Bhandar, Brenna. 2018. *Colonial Lives of Property: Land and Racial Regimes of Ownership.* Duke University Press.

Birrell, Jean. 1987. "Common Rights in the Medieval Forest: Disputes and Conflicts in the Thirteenth Century." *Past & Present* 117 (1): 22–49.

The Black Mediterranean Collective. 2021. *The Black Mediterranean: Bodies, Borders, and Citizenship.* Palgrave Macmillan.

Blanc, Paul David. 2016. *Fake Silk: The Lethal History of Viscose Rayon.* Yale University Press.

Blaser, Mario. 2013. "Notes Towards a Political Ontology of 'Environmental' Conflicts." In *Contested Ecologies: Dialogues in the South on Nature and Knowledge,* edited by Lesley Green, 13–27. Human Sciences Research Council Press.

Blaser, Mario. 2025. *For Emplacement: Political Ontology in Two Acts.* Duke University Press.

Blaser, Mario, and Marisol de la Cadena. 2018. "Introduction: Pluriverse, Proposals for a World of Many Worlds." In *A World of Many Worlds,* edited by Marisol de la Cadena and Mario Blaser, 1–22. Duke University Press.

Blith, Walter. 1652. *The English Improver Improved.* Printed for John Wright.

Boisseron, Bénédicte. 2018. *Afro-Dog: Blackness and the Animal Question.* Columbia University Press.

Bonneuil, Christophe, and Jean-Baptiste Fressoz. 2016. *The Shock of the Anthropocene: The Earth, History and Us.* Verso Books.

Boris, Eileen, and Rhacel Salazar Parreñas, eds. 2010. *Intimate Labors: Culture, Technologies, and the Politics of Care.* Stanford University Press.

Brace, Laura. 2001. "Husbanding the Earth and Hedging Out the Poor." In *Land and Freedom: Law, Property Rights and the British Diaspora,* edited by Andrew Buck, John McLaren, and Nancy Wright, 5–18. Ashgate Publishing.

Brace, Laura. 2004. *The Politics of Property: Labour, Freedom and Belonging.* Edinburgh University Press.

Brancaccio, Francesco, and Carlo Vercellone. 2019. "Birth, Death, and Resurrection of the Issue of the Common: A Historical and Theoretical Perspective." *South Atlantic Quarterly,* 118 (4): 699–709.

Bratich, Jack. 2022. *On Microfascism: Gender, War and Death.* Common Notions Press.

Braun, Bruce. 2004. "Environmental Issues: Writing a More-than-Human Urban Geography." *Progress in Human Geography* 29 (5): 635–50.

Braun, Bruce, and Sarah Whatmore, eds. 2010. *Political Matter: Technoscience, Democracy, and Public Life.* University of Minnesota Press.

Bresnihan, Patrick. 2015. "The More-than-Human Commons: From Commons to Commoning." In *Space, Power, and the Commons: The Struggle for Alternative Futures,* edited by Samuel Kirwan, Leila Dawney, and Julian Brigstocke, 105–24. Routledge.

Briggs, Laura. 2000. "The Race of Hysteria: 'Overcivilization' and the 'Savage' Woman in Late Nineteenth-Century Obstetrics and Gynecology." *American Quarterly* 52 (2): 246–73.

Burr, David. 2001. *The Spiritual Franciscans: From Protest to Persecution in the Century After Saint Francis*. Penn State University Press.

Butler, Judith. 1990. *Gender Trouble: Feminism and the Subversion of Identity*. Routledge.

Butler, Judith. 1997. *The Psychic Life of Power: Theories in Subjection*. Stanford University Press.

Butler, Judith. 2004. *Undoing Gender*. Routledge.

Bynum, Caroline Walker. 2011. *Christian Materiality: An Essay on Religion in Late Medieval Europe*. Zone Books.

Byrd, Jodi. 2011. *The Transit of Empire: Indigenous Critiques of Colonialism*. University of Minnesota Press.

Byrd, Jodi, and Michael Rothberg. 2011. "Between Subalternity and Indigeneity: Critical Categories for Postcolonial Studies." *Interventions* 13 (1): 1–12.

Caffentzis, George, and Monty Neill. 2004. "Preface to 'The New Enclosures: Planetary Class Struggle.'" In *Globalize Liberation: How to Uproot the System and Build a Better World*, edited by David Solnit, 61–72. City Lights Books.

Capone, Nicola. 2020. *Lo spazio e la norma*. Ombre Corte.

The Care Collective. 2020. *The Care Manifesto: The Politics of Interdependence*. Verso.

Carnevale, Francesco, and Alberto Baldasseroni. 1999. *Mal da lavoro: Storia della salute dei lavoratori*. Laterza.

Casalini, Brunella. 2016. "Care e riproduzione sociale. Il rimosso della politica e dell'Economia." *Bollettino telematico di filosofia politica*. Retrieved from https://zenodo.org/records/17011705

Cassata, Francesco. 2011. *Building the New Man: Eugenics, Racial Science, and Genetics in Twentieth-Century Italy*. Central University Press.

Casarino, Cesare, and Antonio Negri. 2008. *In Praise of the Common: A Conversation on Philosophy and Politics*. University of Minnesota Press.

Castree, Noel. 1995. "The Nature of Produced Nature: Materiality and Knowledge Construction in Marxism." *Antipode* 27 (1): 12–48.

Cellamare, Carlo. 2014. "The Self-Made City." In *Global Rome: Changing Faces of the Eternal City*, edited by Isabella Clough Marinaro and Bjorn Thomassen, 205–18. Indiana University Press.

Centemeri, Laura. 2018. "Commons and the New Environmentalism of Everyday Life: Alternative Value Practices and Multispecies Commoning in the Permaculture Movement." *Rassegna italiana di Sociologia* 59 (2): 289–314.

Cerretano, Valerio. 2018. "Multinational Business and Host Countries in Times of Crisis: Courtaulds, Glanzstoff, and Italy in the Inter-War Period." *Economic History Review* 71 (2): 540–66.

Césaire, Aimé. 1972. *Discourse on Colonialism*. Monthly Review Press.

Chakrabarty, Dipesh. 2009. "The Climate of History: Four Theses." *Critical Inquiry* 35 (2): 197–222.

Charbonnier, Pierre. 2021. *Affluence and Freedom: An Environmental History of Political Ideas*. Polity Press.

Chen, Mel Y. 2012. *Animacies: Biopolitics, Racial Mattering, and Queer Affect*. Duke University Press.

Cielo, Cristina, and Lisset Cabo. 2018. "Extractivism, Gender, and Disease: An Intersectional Approach to Inequalities." *Ethics and International Affairs* 32 (2): 169–78.

Cielemcka, Olga, and Cecilia Åsberg. 2019. "Introduction: Toxic Embodiment and Feminist Environmental Humanities." *Environmental Humanities* 11 (1): 101–7.

Cirillo, Valeria. 2024. "Abitare le catastrofi ecologiche. Prospettive multispecie a partire dalla filosofia speculative di Alfred North Whitehead." PhD dissertation. Dept. of Philosophy, Roma Tre University.

Clement, Floriane, Wendy Harcourt, Deepa Joshi, and Chizu Sato. 2019. "Feminist Political Ecologies of the Commons and Commoning." *International Journal of the Commons* 13 (1): 1–15.

Cohen, Ed. 2009. *A Body Worth Defending: Immunity, Biopolitics, and the Apotheosis of the Modern Body*. Duke University Press.

Cohen, Ed. 2013. "Human Tendencies." *e-misférica* 10 (1). Retrieved from https://hemisphericinstitute.org/en/emisferica-101/10-1-essays/human-tendencies.html.

Cohen, Jeffrey Jerome. 2015. *Stone: An Ecology of the Inhuman*. University of Minnesota Press.

Cohn, Norman. 1970. *The Pursuit of the Millennium: Revolutionary and Mystical Anarchists in the Middle Ages*. Oxford University Press.

Colectivo Miradas Criticas. 2017. *Mapeando el Cuerpo-territorio. Guía metodológica Para Mujeres que Defienden sus Territorios*. https://miradascriticasdelterritoriodesdeelfeminismo.files.wordpress.com/2017/11/mapeando-el-cuerpo-territorio.pdf.

Coleman, Janet. 2011. "Medieval Political Theory c.1000–1500." In *The Oxford Handbook of the History of Political Philosophy*, edited by George Klosko, 180–205. Oxford University Press.

Collard, Rosemary-Claire, and Jessica Dempsey. 2018. "Accumulation by Difference-Making: An Anthropocene Story, Starring Witches." *Gender, Place & Culture* 25 (9): 1349–64.

Collard, Rosemary-Claire, Jessica Dempsey, and Juanita Sundberg. 2015. "A Manifesto for Abundant Futures." *Annals of the Association of American Geographers* 105 (2): 322–30.

Combes, Muriel. 2013. *Gilbert Simondon and the Philosophy of the Transindividual*. MIT Press.

Cooper, Melinda. 2010. "Turbulent Worlds." *Theory, Culture & Society* 27 (2–3): 167–90.

Cooper, Melinda, and Catherine Waldby. 2014. *Clinical Labor: Tissue Donors and Research Subjects in the Global Bioeconomy*. Duke University Press.

Coulthard, Glen. 2010. "Place Against Empire: Understanding Indigenous Anti-Colonialism." *Affinities: A Journal of Radical Theory, Culture, and Action* 4 (2): 79–83.

Coulthard, Glen. 2014. *Red Skin, White Masks: Rejecting the Colonial Politics of Recognition*. University of Minnesota Press.

Cronon, William. 2003. *Changes in the Land: Indians, Colonists, and the Ecology of New England*. Hill & Wang.

Cruz, Delmy Tania, Manuel Bayon Jiménez, and Colectivo Miradas Criticas del Territorio desde el Feminismo. 2022. *Bodies, Territories, and Feminisms: Latin American Compilation of Political Practices, Theories, and Methodologies*. Ibidem.

Cusicanqui, Silvia Rivera. 2015. "Strategic Ethnicity, Nation, and (Neo)colonialism in Latin America." *Alternautas* 2 (2): 10–20.

Dalla Costa, Mariarosa. 2005. "Mariarosa Dalla Costa." In *Gli operaisti: autobiografie dei cattivi maestri*, edited by Guido Borio, Francesca Pozzi, and Gigi Roggero, 121–40. Derive Aprodi.

Dalla Costa, Mariarosa. 2019. *Women and the Subversion of the Community: A Mariarosa Dalla Costa Reader*. PM Press.

Dalla Costa, Mariarosa, and Monica Chilese. 2014. *Our Mother Ocean: Enclosures, Commons and the Global Fishermen's Movement*. Commons Notions.

Danowski, Déborah, and Eduardo Viveiros de Castro. 2016. *The Ends of the World*. Polity Press.

Dardot, Pierre, and Christian Laval. 2015. *Common: On Revolution in the 21st Century*. Bloomsbury.

Daston, Lorraine, and Katherine Park. 2001. *Wonders and the Order of Nature: 1150–1750*. Zone Books.

Davis, Angela. Y. 1981. *Women, Race, & Class*. Vintage Books.

De Angelis, Massimo. 2007. *The Beginnings of History: Value Struggles and Global Capital*. Pluto Press.

De Angelis, Massimo. 2017. *Omnia Sunt Communia: On the Commons and the Transformation to Postcapitalism*. Bloomsbury Publishing.

De Angelis, Massimo, and Dagmar Diesner. 2020. "A Revolution Under Our Feet: Food Sovereignty and the Commons in the Case of Campi Aperti." In *Commoning the City*, edited by Derya Özkan and Güldem Baykal Büyüksaraç, 69–85. Routledge.

De la Bellacasa, Maria Puig. 2017. *Matters of Care: Speculative Ethics in More than Human Worlds*. University of Minnesota Press.

De la Cadena, Marisol. 2010. "Indigenous Cosmopolitics in the Andes: Conceptual Reflections Beyond 'Politics.'" *Cultural Anthropology* 25 (2): 334–70.

De la Cadena, Marisol. 2013. "About 'Mariano's Archive': Ecologies of Stories." In *Contested Ecologies: Dialogues in the South on Nature and Knowledge*, edited by Lesley Green, 55–68. Human Sciences Research Council Press.

De la Cadena, Marisol. 2015a. *Earth Beings: Ecologies of Practice Across Andean Worlds*. Duke University Press.

De la Cadena Marisol. 2015b. "Uncommoning Nature." *e-flux* Supercommunity. http://supercommunity.e-flux.com/texts/uncommoning-nature/.

De Palma, Vittoria. 2014. *Wasteland: A History*. Yale University Press.

Deleuze, Gilles. 1988. *Foucault*. University of Minnesota Press.

Deleuze, Gilles. 2004. *Desert Islands: And Other Texts 1953–1974*. Semiotext(e).

Deleuze, Gilles, and Felix Guattari. 1987. *A Thousand Plateaus: Capitalism and Schizophrenia*. Continuum.

Derrida, Jacques. 1996. *Archive Fever: A Freudian Impression*. University of Chicago Press.

Deslandes, Anne Louise. 2024. "Crossing the Storm: EZLN Marks 30 Years with a 120-Year Plan." *NACLA Report on the Americas*. Retrieved from https://nacla.org/crossing-storm-ezln-marks-30-years-120-year-plan.

Di Chiro, Giovanna. 2017. "Welcome to the White (M)anthropocene?: A Feminist-Environmentalist Critique." In *Routledge Handbook of Gender and Environment*, edited by Sherilyn MacGregor, 487–505. Routledge.

Di Feliciantonio, Cesare, and Silvia Aru. 2018. "From (Urban) Commons to Commoning: Political Practices and Horizons in the Mediterranean Context. Introduction to the Special Issue." *Acme: An International Journal for Critical Geographies* 17 (2): 258–68.

Dowling, Emma. 2020. *The Care Crisis*. Verso Books.

Ehrlich, Paul. 1968. *The Population Bomb*. Ballantine Books.

Eidelman, Tessa A., and Sara Safransky. 2020. "The Urban Commons: A Keyword Essay." *Urban Geography* 42 (6): 792–811.

Ellis, Erle. 2012. "The Planet of No Return: Human Resilience on an Artificial Earth." *Breakthrough Journal*, 2: 37–44.

Enright, Theresa, and Ugo Rossi. 2018. "Ambivalence of the Urban Commons." In *The Routledge Handbook on Spaces of Urban Politics*, edited by Kevin Ward, Andrew E. G. Jones, Byron Miller, and David Wilson, 35–46. Routledge.

Ernstson, Henrik, and Erik Swyngedouw. 2019. "Politicizing the Environment in the Urban Century." In *Urban Political Ecology in the Anthropo-Obscene. Interruptions and Possibilities*, edited by Henrik Ernstson and Erik Swyngedouw, 3–21. Routledge.

Escobar, Arturo. 2020. *Pluriversal Politics: The Real and the Possible*. Duke University Press.

Esposito, Roberto. 2015. *Persons and Things: From the Body's Point of View*. Polity Press.

Esposito, Roberto. 2022. *Institution*. Polity Press.

Estes, Nick. 2019. *Our History Is the Future: Standing Rock Versus the Dakota Access Pipeline and the Long Tradition of Indigenous Resistance*. Verso Books.

Esteva, Gustavo. 2004. *Celebration of Zapatismo*. Multiversity.

Fabricant, Nicole, and Bret Gustafson. 2016. "Revolutionary Extraction? Mapping the Political Economy of Gas, Soy, and Mineral Production in Evo Morales's Bolivia." *NACLA Report on the Americas* 48 (3): 271–79.

Farris, Sara R. 2022. *Giorgia Meloni is a Female Face for an Anti-Feminist Agenda*. Jacobin. Retrieved from https://jacobin.com/2022/12/giorgia-meloni-far-right-feminism-nationalism-family.

Farvadin, Firoozeh, and Gustavo Robles. 2025. "From (Individual) Fears to (Collective) Cares." *Irgac.org*. Retrieved from https://irgac.org/articles/from-individual-fears-to-collective-cares/.

Federici, Silvia. 1975. *Wages Against Housework*. Falling Wall Press.

Federici, Silvia. 2004. *Caliban and the Witch: Women, the Body, and Primitive Accumulation*. Autonomedia.

Federici, Silvia. 2010. "Feminism and the Politics of the Commons in an Era of Primitive Accumulation." In *Uses of a Whirlwind: Movement, Movements, and Contemporary Radical Currents in the United States*, edited by Team Colors Collective, 274–94. AK Press.

Federici, Silvia. 2018. *Re-Enchanting the World: Feminism and the Politics of the Commons*. PM Press.
Federici, Silvia. 2020. "Silvia Federici." In *Revolutionary Feminisms: Conversations on Collective Action and Radical Thought*, edited by Brenna Bhandar and Rafeef Ziadah, 149–57. Verso Books.
Feltrin, Lorenzo, and Devi Sacchetto. 2021. "The Work-Technology Nexus and Working-Class Environmentalism: Workerism Versus Capitalist Noxiousness in Italy's Long 1968." *Theory and Society* 50 (5): 815–35.
Ferdinand, Malcolm. 2021. *Decolonial Ecology. Thinking from the Caribbean World*. Polity Press.
Fortier, Craig. 2017. *Unsettling the Commons. Social Movements Within, Against, and Beyond Settler Colonialism*. Arp Books.
Fortunati, Leopoldina. 1981. *L'arcano della riproduzione: casalinghe, prostitute, operai e capitale*. Marsilio.
Foster, John Bellamy. 2000. *Marx's Ecology: Materialism and Nature*. New York University Press.
Foucault, Michel. 1970. *The Order of Things: An Archaeology of the Human Sciences*. Routledge.
Foucault, Michel. 1995. *Discipline and Punish: The Birth of the Prison*. Vintage.
Foucault, Michel. 2003. *Society Must be Defended: Lectures at the Collège de France 1975–1976*. Picador.
Foucault, Michel. 2007. *Security, Territory, Population: Lectures at the Collège de France 1977–78*. Springer.
Fox Keller, Evelyn. 1985. *Reflections on Gender and Science*. Yale University Press.
Fragnito, Maddalena, and Miriam Tola. 2021. "Nella zona nevralgica del conflitto: note su femminismi e cura." In *Ecologie della cura. Prospettive transfemministe*, edited by Maddalena Fragnito and Miriam Tola, 7–28. Orthotes.
Fraser, Nancy. 2016. "Contradictions of Capital and Care." *New Left Review* 100: 99–117.
Gaard, Greta. 2011. "Ecofeminism Revisited: Rejecting Essentialism and Re-Placing Species in a Material Feminist Environmentalism." *Feminist formations* 23 (2): 26–53.
Gago, Verónica. 2020. *The Feminist International: How to Change Everything*. Verso Books.
Gago, Verónica, 2025. "The Reinvention of the Strike: 10 Years of Feminist Uprisings in Argentina." *Ojala.mx*. Retrieved from https://www.ojala.mx/en/ojala-en/the-reinvention-of-the-strike-as-a-tool-10-years-of-feminist-uprising-in-argentina
Gago, Verónica, and Sandro Mezzadra. 2017. "A Critique of the Extractive Operations of Capital: Toward an Expanded Concept of Extractivism." *Rethinking Marxism* 29 (4): 574–91.
Gago, Verónica, Raquel Gutiérrez Aguilar, Susana Draper, Mariana Menéndez Díaz, Marina Montanelli, and Suely Rolnik. 2018. *8M Constelación feminista: ¿cuál es tu huelga?: ¿cuál es tu lucha?*. Tinta Limón.
García López, Gustavo A., Irina Velicu, and Giacomo D'Alisa. 2017. "Performing Counter-Hegemonic Common (s) Senses: Rearticulating Democracy, Community, and Forests in Puerto Rico." *Capitalism Nature Socialism* 28 (3): 88–107.

Garnsey, Peter. 2007. *Thinking About Property: From Antiquity to the Age of Revolution.* Cambridge University Press.

Gibson-Graham, J. K., and Gerda Roelvink. 2010. "An Economic Ethics for the Anthropocene." *Antipode* 41 (1): 320–46.

Gibson-Graham, J. K., Jenny Cameron, and Stephen Healy. 2016. "Commoning as a Postcapitalist Politics." In *Releasing the Commons: Rethinking the Futures of the Commons*, edited by Ash Amin and Philip Howell, 192–212. Routledge.

Ginzburg, Carlo. 1992. *The Cheese and the Worms: The Cosmos of a Sixteenth-Century Miller.* Johns Hopkins University Press.

Giuliani, Alfonso, and Carlo Vercellone. 2019. "From New Institutional Economics of the Commons to the Common as a Mode of Production." *South Atlantic Quarterly* 118 (4): 767–87.

Goldman, Michael, ed. 1998. *Privatizing Nature: Political Struggles for the Global Commons.* Pluto Press.

Gómez-Barris, Macarena. 2017. *The Extractive Zone: Social Ecologies and Decolonial Perspectives.* Duke University Press.

Graeber, David. 2013. *The Democracy Project: A History, a Crisis, a Movement.* Random House.

Greer, Allan. 2012. "Commons and Enclosure in the Colonization of North America." *The American Historical Review* 117 (2): 365–86.

Greer, Allan. 2018. *Property and Dispossession: Natives, Empires, and Land in Early Modern North America.* Cambridge University Press.

Grossi, Paolo. 1972. "Usus facti: La nozione di proprietà nella inaugurazione dell'età nuova." *Quaderni fiorentini per la storia del pensiero giuridico moderno* 1 (1): 287–355.

Grosz, Elizabeth. 2012. "Identity and Individuation: Some Feminist Reflections." In *Gilbert Simondon: Being and Technology*, edited by Arne De Boever et al., 37–56. Edinburgh University Press.

Hamilton, Jennifer, and Astrida Neimanis. 2018. "Composting Feminisms and Environmental Humanities." *Environmental Humanities* 10 (2): 501–27.

Hanawalt, Barbara, and Lisa Kiser. 2008. *Engaging with Nature: Essays on the Natural World in Medieval and Early Modern Europe.* University of Notre Dame Press.

Haraway, Donna J. 1989. *Primate Visions: Gender, Race, and Nature in the World of Modern Science.* Routledge.

Haraway, Donna J. 2008. *When Species Meet.* University of Minnesota Press.

Haraway, Donna J. 2016. "Staying with the Trouble: Anthropocene, Capitalocene, Chthulucene." In *Anthropocene or Capitalocene? Nature, History, and the Crisis of Capitalism*, edited by Jason W. Moore, 34–76. PM Press.

Hardin, Garrett. 1968. "The Tragedy of the Commons: The Population Problem Has No Technical Solution; It Requires a Fundamental Extension in Morality." *Science* 162 (3859): 1243–48.

Hardt, Michael. 2010. "Two Faces of Apocalypse: A Letter from Copenhagen." *Polygraph* 22: 265–74.

Hardt, Michael, and Antonio Negri. 2000. *Empire.* Harvard University Press.

Hardt, Michael, and Antonio Negri. 2009. *Commonwealth.* Harvard University Press.

Hardt, Michael, and Antonio Negri. 2019. "Empire, Twenty Years On." *New Left Review* 120: 67–92.

Harney, Stefano, and Fred Moten. 2021. *All Incomplete.* Minor Compositions.

Harris, Cheryl I. 1993. "Whiteness as Property." *Harvard Law Review* 106 (8): 1707–91.

Hartman, Saidiya. 1997. *Scenes of Subjection: Terror, Slavery, and Self-Making in Nineteenth-Century America.* Oxford University Press.

Harvey, David. 1991. *The Condition of Postmodernity: An Enquiry into the Origins of Cultural Change.* Wiley-Blackwell.

Harvey, David. 2000. *Spaces of Hope.* University of California Press.

Harvey, David. 2003. *The New Imperialism.* Oxford University Press.

Harvey, David. 2012. *Rebel Cities: From the Right to the City to the Urban Revolution.* Verso Books.

Harvie, David, and Ben Trott. 2021. "Cognitive Capitalism." In *The SAGE Handbook of Marxism,* edited by Beverley Skeggs, Sara R. Farris, Alberto Toscano, and Svenja Bromberg, 1517–38. SAGE.

Heynen, Nik. 2021. "A Plantation Can Be a Commons: Re-Earthing Sapelo Island Through Abolition Ecology: The 2018 Neil Smith Lecture." *Antipode* 53 (1): 95–114.

Heynen Nik, Maria Kaika, and Erik Swyngedouw, eds. 2006. *In the Nature of Cities. Urban Political Ecology and the Politics of Urban Metabolism.* Routledge.

Hindery, Derrick. 2013. *From Enron to Evo: Pipeline Politics, Global Environmentalism, and Indigenous Rights in Bolivia.* University of Arizona Press.

Hobart, Hi'ilei Julia Kawehipuaakahaopulan, and Tamara Kneese. 2020. "Radical Care: Survival Strategies for Uncertain Times." *Social Text* 38 (1): 1–16.

Hoffmann, Richard. 2014. *An Environmental History of Medieval Europe.* Cambridge University Press.

Huber, Matthew. 2013. *Lifeblood: Oil, Freedom, and the Forces of Capital.* University of Minnesota Press.

Huron, Amanda. 2018. *Carving Out the Commons: Tenant Organizing and Housing Cooperatives in Washington.* University of Minnesota Press.

Ince, Onur Ulas. 2018. *Colonial Capitalism and the Dilemmas of Liberalism.* Oxford University Press.

Insolera, Italo. 1962. *Roma moderna: un secolo di storia urbanistica.* Einaudi.

Jackson, Steven J. 2014. "Rethinking Repair." In *Media Technologies: Essays on Communication, Materiality, and Society,* 2, edited by Tarleton Gillespie, Pablo J. Boczkowski, and Kirsten A. Foot, 21–39. MIT Press.

Jackson, Zakiyya Iman. 2020. *Becoming Human: Matter and Meaning in an Antiblack World.* New York University Press.

Johnson, Elizabeth R., and Jesse Goldstein. 2015. "Biomimetic Futures: Life, Death, and the Enclosure of a More-than-Human Intellect." *Annals of the Association of American Geographers* 105 (2): 387–96.

Kaika, Maria. 2017. "'Don't Call Me Resilient Again!': The New Urban Agenda as Immunology . . . or . . . What Happens When Communities Refuse to be Vaccinated with 'Smart Cities' and Indicators." *Environment and Urbanization* 29 (1): 89–102.

Karaman, Ozan. 2013. "Defending Future Commons: The Gezi Experience." *Antipode.org*. Retrieved from https://antipodeonline.org/2013/08/27/intervention-defending-future-commons-the-gezi-experience-by-ozan-karaman-2/.

Kauanui, J. Kēhaulani. 2015. "Nothing Common About 'the Commons': Settler Colonialism and Indigenous Difference." Unpublished paper.

Kennedy, Rosanne. 2014. "Multidirectional Eco-Memory in an Era of Extinction: Colonial Whaling and Indigenous Dispossession in Kim Scott's 'The Deadman Dance.'" In *The Routledge Companion to Environmental Humanities*, edited by Ursula Heise, Jon Christensen, and Michelle Niemann, 268–77. Routledge.

Khasnabish, Alex. 2008. *Zapatismo Beyond Borders: New Imaginations of Political Possibility*. University of Toronto Press.

Kiser, Lisa. 2003. "The Garden of St. Francis: Plants, Landscape, and Economy in Thirteenth-Century Italy." *Environmental History* 8 (2): 229–45.

Kohn, Margaret, and Keally McBride. 2011. *Political Theories of Decolonization*. Oxford University Press.

Labban, Mazen. 2014. "Deterritorializing Extraction: Bioaccumulation and the Planetary Mine." *Annals of the Association of American Geographers* 104 (3): 560–76.

Lancione, Michele. 2024. *For a Liberatory Politics of Home*. Duke University Press.

Law, John. 2015. "What's Wrong with a One-World World?" *Distinktion* 16 (1): 126–39.

Leff, Gordon. 1999. *Heresy in the Later Middle Ages: The Relation of Heterodoxy to Dissent c.1250–c.1450*. Manchester University Press.

LeMenager, Stephanie. 2021. "The Commons." In *The Cambridge Companion to Environmental Humanities*, edited by Jeffrey Cohen and Stephanie Foote, 11–25. Cambridge University Press.

Lenkersdorf, Carlos. 1996. *Los hombres verdaderos: Voce y testimonies tojolabales*. Siglo Veintiuno.

Leonardi, Emanuele. 2017. *Lavoro, natura, valore. André Gorz tra marxismo e decrescita*. Orthotes.

Leonardi, Emanuele. 2019. "Bringing Class Analysis Back In: Assessing the Transformation of the Value-Nature Nexus to Strengthen the Connection Between Degrowth and Environmental Justice." *Ecological Economics* 156 (2019): 83–90.

Leonardi, Emanuele. 2021. "Autonomist Marxism and World-Ecology: For a Political Theory of the Ecological Crisis." *PPPR*. https://projectpppr.org/pandemics/autonomist-marxism-and-world-ecology-for-a-political-theory-of-the-ecological-crisis.

Liboiron, Max. 2021. *Pollution is Colonialism*. Duke University Press.

Linebaugh, Peter. 2008. *The Magna Carta Manifesto: Commons and Liberties for All*. University of California Press.

Linebaugh, Peter, and Marcus Rediker. 2000. *The Many-Headed Hydra: Sailors, Slaves, Commoners, and the Hidden Histories of the Revolutionary Atlantic*. Beacon Press.

Livingston, Julie. 2019. *Self-Devouring Growth: A Planetary Parable as Told from Southern Africa*. Duke University Press.
Locke, John. 1988. *Two Treatises of Government*. Cambridge University Press.
Lorde, Audre. 1988. *A Burst of Light: Essays*. Firebrand Books.
Loriga, Giovanni. 1925. "Le condizioni igieniche nell'industria della seta artificiale." *Bollettino del lavoro e della previdenza sociale* 5: 86–95.
Lorey, Isabell. 2015. *State of Insecurity: Government of the Precarious*. Verso Books.
Lotringer, Sylvere, and Christian Marazzi. 1980. *Italy: Autonomia, Post-Political Politics*. Semiotext(e).
Lowe, Lisa. 2015. *The Intimacies of Four Continents*. Duke University Press.
Lowe, Lisa. 2020. "Afterword: Revolutionary Feminisms In a Time of Monsters." In *Revolutionary Feminisms: Conversations on Collective Action and Radical Thought*, edited by Brenna Bhandar and Rafeef Ziadah, 217–27. Verso Books.
Lucha y Siesta. 2022. *Dichiarazione di autogoverno*. Lucha Y Siesta. Retrieved from https://luchaysiesta.org/la-dichiarazione-di-autogoverno/.
Lugones, Maria. 2007. "Heterosexualism and the Colonial/Modern Gender System." *Hypatia* 22 (1): 186–219.
Luisetti, Federico, John Pickles, and Wilson Kaiser. 2015. *The Anomie of the Earth. Philosophy, Politics, and Autonomy in Europe and the Americas*. Duke University Press.
Mackenzie, Adrian. 2002. *Transductions: Bodies and Machines at Speed*. Continuum.
MacGregor, Sherilyn. 2006. *Beyond Mothering Earth: Ecological Citizenship and the Politics of Care*. University of British Columbia Press.
Macpherson, Crawford Brough. 1962. *The Political Theory of Possessive Individualism: Hobbes to Locke*. Oxford University Press.
Mäkinen, Virpi. 2001. *Property Rights in the Late Medieval Discussion on Franciscan Poverty*. Peeters.
Malatino, Hil. 2020. *Trans Care*. University of Minnesota Press.
Marcos, Sylvia. 2014. "The Zapatista Women's Revolutionary Law As It Is Lived Today." *Open Democracy*. Retrieved from http://www.opendemocracy.net/sylviamarcos/zapatista-women%E2%80%99s-revolutionary-law-as-it-is-lived-today.
Marcos, Sylvia. 2021. "Las mujeres Zapatistas reconceptualizan su lucha." *Tabula Rasa* 38: 197–211.
Manjapra, Kris. 2020. *Colonialism in Global Perspective*. Cambridge University Press.
Markham, Gervase. 1986. *The English Housewife*. Edited by Michael R. Best. McGill-Queen's University Press.
Marx, Karl. 1973. *Grundrisse: Foundations of the Critique of Political Economy*. Penguin.
Marx, Karl. 1976. *Capital: Volume 1*. Penguin.
Marx, Karl. 1988. *The Economic and Philosophic Manuscripts of 1844 and the Communist Manifesto*. Great Books in Philosophy.
Marx, Karl, and Friedrich Engels. 1976. *Manifesto of the Communist Party*. Marx and Engels Collected Works, Vol. 6. Lawrence & Wishart.
Massumi, Brian. 2018. *99 Theses on the Revaluation of Value: A Postcapitalist Manifesto*. University of Minnesota Press.

Mattei, Ugo. 2011. *Beni comuni. Un manifesto*. Laterza.
Mayor Ana Maria. 1996. "Discurso inaugural de la Mayor Ana Maria en el primero encuentro intercontinental por la humanidad y contra el neoliberalismo." *Chiapas* 3: 101–5.
Maxwell, Alexander. 2014. *Patriots Against Fashion: Clothing and Nationalism in Europe's Age of Revolutions*. Palgrave Macmillan.
Mbembe, Achille. 2001. *On the Postcolony*. University of California Press.
Mbembe, Achille. 2021. "The Universal Right to Breathe." *Critical Inquiry* 47 (2): 58–62.
McClintock, Anne. 1995. *Imperial Leather: Race, Gender, and Sexuality in the Colonial Contest*. Routledge.
McClure, Julia. 2017. *The Franciscan Invention of the New World*. Palgrave Macmillan.
McCormick, Ted. 2009. *William Petty: And the Ambitions of Political Arithmetic*. Oxford University Press.
Mehta, Uday Singh. 1999. *Liberalism and Empire: A Study in Nineteenth-Century British Liberal Thought*. University of Chicago Press.
Mentinis, Mihalis. 2006. *Zapatistas: The Chiapas Revolt and What It Means for Radical Politics*. Pluto Press.
Merchant, Caroline. 1980. *The Death of Nature: Women, Ecology, and the Scientific Revolution*. HarperCollins.
Merchant, Caroline. 2003. *Reinventing Eden: The Fate of Nature in Western Culture*. Routledge.
Meziane, Mohamed Amer. 2024. *The States of the Earth. An Ecological and Racial History of Secularization*. Verso Books.
Mezzadra, Sandro, and Brett Neilson. 2013. *Border As Method; Or, the Multiplication of Labor*. Duke University Press.
Mezzadra, Sandro, and Brett Neilson. 2019. *The Politics of Operations: Excavating Contemporary Capitalism*. Duke University Press.
Mezzadri, Alessandra. 2019. "On the Value of Social Reproduction: Informal Labour, the Majority World and the Need for Inclusive Theories and Politics." *Radical Philosophy* 2 (4): 33–41.
Mezzadri, Alessandra. 2022. "Social Reproduction and Pandemic Neoliberalism: Planetary Crises and the Reorganisation of Life, Work and Death." *Organization* 29 (3): 379–400.
Midnight Notes Collective. 1990. "The New Enclosures." *Midnight Notes* 10: 1–9.
Mies, Maria. 1986. *Patriarchy and Accumulation on a World Scale: Women in the International Division of Labour*. Zed Books.
Mies, Maria, and Vandana Shiva. 1993. *Ecofeminism*. Zed Books.
Mies, Maria, and Veronika Bennholdt-Thomsen. 1999. *The Subsistence Perspective: Beyond the Globalized Economy*. Zed Books.
Mignolo, Walter. 2011. *The Darker Side of Western Modernity: Global Futures, Decolonial Options*. Duke University Press.
Mignolo, Walter, and Freya Schiwy. 2002. "Translation/Transculturation and the Colonial Difference." In *Beyond Dichotomies: Histories, Identities, Cultures, and the*

Challenge of Globalization, edited by Elisabeth Mudimbe Boyi, 251–86. Syracuse University Press.

Molavi, Shourideh. 2024. *Environmental Warfare in Gaza*. Pluto Press.

Mollett, Sharlene, and Caroline Faria. 2013. "Messing with Gender in Feminist Political Ecology." *Geoforum* 45: 116–25.

Montanelli, Marina. 2018. "The Unforeseen Subject of the Feminist Strike." *South Atlantic Quarterly* 117 (3): 699–709.

Monticelli, Lara, ed. 2022. *The Future Is Now: An Introduction to Prefigurative Politics*. Bristol University Press.

Moore, Jason. 2015. *Capitalism in the Web of Life: Ecology and the Accumulation of Capital*. Verso Books.

Moore, Jason. 2018. "The Capitalocene Part II: Accumulation by Appropriation and the Centrality of Unpaid Work/Energy." *Journal of Peasant Studies* 45 (2): 237–79.

Moulier-Boutang, Yann. 2012. *Cognitive Capitalism*. Polity Press.

Mudu, Pierpaolo. 2004. "Resisting and Challenging Neoliberalism: Development of Italian Social Centers." *Antipode* 36 (5): 917–41.

Muecke, Stephen. 2017. "Attachment." *Environmental Humanities* 9 (1): 167–70.

Mullings, Beverley. 2021. "Caliban, Social Reproduction and Our Future Yet to Come." *Geoforum* 118: 150–58.

Muñoz, José Esteban. 2018. "Preface: Fragment from the *Sense of Brown* Manuscript." *GLQ. A Journal of Lesbian and Gay Studies* 24 (4): 395–97.

Munroe, Jennifer. 2008. *Gender and the Garden in Early Modern English Literature*. Ashgate Publishing.

Murphy, Michelle. 2008. "Chemical Regimes of Living." *Environmental History* 13 (4): 695–703.

Murphy, Michelle. 2010. "Chemical Infrastructures of the St. Clair River." In *Toxicants, Health and Regulation Since 1945*, edited by Soraya Boudia and Nathalie Jas, 103–15. Pickering & Chatto.

Murphy, Michelle. 2013. "Distributed Reproduction, Chemical Violence, and Latency." *Scholar and Feminist Online* 11 (3). Retrieved from https://sfonline.barnard.edu/distributed-reproduction-chemical-violence-and-latency/.

Neeson, Jeanette M. 1996. *Commoners: Common Right, Enclosure, and Social Change in England 1700–1820*. Cambridge University Press.

Negri, Antonio. 1999. *Insurgencies: Constituent Power and the Modern State*. University of Minnesota Press.

Negri, Antonio. 2010. *Diary of an Escape*. Polity Press.

Neimanis, Astrida. 2021. "An Archive of an Epoch that Almost Was." In *Feminist, Queer, Anticolonial Propositions for Hacking the Anthropocene: Archive*, edited by Jennifer Mae Hamilton, Susan Reid, Pia van Gelder, and Astrida Neimanis, 7–13. Open Humanities Press.

Nelson, Sara. 2014. "Resilience and the Neoliberal Counter-Revolution: From Ecologies of Control to Production of the Common." *Resilience* 2 (1): 1–17.

Nelson, Sara, and Bruce Braun. 2017. "Autonomia in the Anthropocene: New Challenges to Radical Politics." *South Atlantic Quarterly* 116 (2): 223–35.

Nerbini, Gino. 1925. "Industrie romane : L'industria della seta artificiale." *Capitolium* 3: 160–63.

Neyrat, Frédéric. 2019. *The Unconstructable Earth: An Ecology of Separation*. Fordham University Press.

Nixon, Rob. 2011. *Slow Violence and the Environmentalism of the Poor*. Harvard University Press.

Nixon, Rob. 2012. "Neoliberalism, Genre, and 'The Tragedy of the Commons.'" *PMLA* 127 (3): 593–99.

Non Una di Meno Roma. 2020. "Life Beyond the Pandemic." *Interface: A Journal for and About Social Movements* 12 (1): 109–14.

Olivera, Marcela. 2015. "Water Beyond the State." In *Patterns of Commoning*, edited by David Bollier and Silke Helfrich, 86–91. Levellers Press.

Olivera, Marcela. 2019. "Working with the Commons." *e-flux Architecture*. Retrieved from https://www.e-flux.com/architecture/liquid-utility/259666/working-with-the-commons/.

Olwig, Kenneth R. 2016. "Performing on the Landscape Versus Doing Landscape: Perambulatory Practice, Sight and the Sense of Belonging." In *Ways of Walking: Ethnography and Practice on Foot*, edited by Tim Ingold and Jo Lee Vergunst, 93–104. Routledge.

Osborne, Peter. 2005. *How to Read Marx*. Granta.

Osco, Marcelo F. 2010. "Ayllu: Critical Thinking and (An)other Autonomy." In *Indigenous Peoples and Autonomy: Insights for a Global Age*, edited by Mario Blaser, Ravi de Costa, Deborah McGregor, and William D. Coleman, 27–48. University of British Columbia Press.

Ostrom, Elinor. 1990. *Governing the Commons: The Evolution of Institutions for Collective Action*. Cambridge University Press.

Papadopoulos, Dimitris. 2018. *Experimental Practice: Technoscience, Alterontologies, and More-than-Social Movements*. Duke University Press.

Papadopoulos, Dimitris, Maria Puig de la Bellacasa, and Maddalena Tacchetti, eds. 2023. *Ecological Reparation. Repair, Remediation and Resurgence in Social and Environmental Conflict*. Bristol University Press.

Pasolini, Pier Paolo. 2016. *The Street Kids*. Europa Editions.

Pellizzoni, Luigi. 2015. *Ontological Politics in a Disposable World: The New Mastery of Nature*. Ashgate Publishing.

Pellizzoni, Luigi. 2025. *Nature, Neoliberalism, and New Materialisms: Riding the Ungovernable*. Lexington Books.

Pesarini, Angelica. 2023. "Biopolitiche di razza, genere e cittadinanza nel discorso politico italiano." *From the European South: A Transdisciplinary Journal of Postcolonial Humanities* 12 (2023): 39–58.

Petersmann, Marie. 2024. "Becoming Common: Ecological Resistance, Refusal, Reparation." In *International Law and Posthuman Theory,* edited by Matilda Arvidsson and Emily Jones, 222–43. Routledge.

Petty, William Sir. 1899. *The Economic Writings of Sir William Petty: Together with the Observations Upon the Bills of Mortality.* Cambridge University Press.

Pignarre, Philippe, and Isabelle Stengers. 2011. *Capitalist Sorcery: Breaking the Spell.* Polity Press.

Pinkus, Karen. 1995. *Bodily Regimes: Italian Advertising Under Fascism.* University of Minnesota Press.

Pinto, Isabella, Chiara Belingardi, Ilenia Caleo, and Federica Giardini. 2014. "Spatial Struggles: Teatro Valle Occupato and the (Right to the) City." *Open Democracy.* Retrieved from https://www.opendemocracy.net/en/can-europe-make-it/spatial-struggles/

Pirtle, Whitney N. Laster. 2020. "Racial Capitalism: A Fundamental Cause of Novel Coronavirus (COVID-19)." *Health Education & Behavior* 47 (4): 504–8.

Plumwood, Val. 1993. *Feminism and the Mastery of Nature.* Routledge.

Posthumus, Stephanie. 2017. *French "Ecocritique": Reading Contemporary French Theory and Fiction Ecologically.* University of Toronto Press.

Povinelli, Elizabeth. 2012. "The will to be otherwise / The effort of endurance." *The South Atlantic Quarterly* 111 (3): 453–75.

Povinelli, Elizabeth. 2016. *Geontologies: A Requiem to Late Liberalism.* Duke University Press.

Prigogine, Ilya, and Isabelle Stengers. 2018. *Order Out of Chaos: Man's New Dialogue with Nature.* Verso Books.

Pulido, Laura. 2018. "Racism and the Anthropocene." In *Future Remains: A Cabinet of Curiosities for the Anthropocene,* edited by Gregg Mitman, Marco Armiero, and Robert Emmett, 116–28. University of Chicago Press.

Raghuram, Parvati. 2021. "Race and Feminist Care Ethics: Intersectionality as Method." In *The Changing Ethos of Human Rights,* edited by Hoda Mahmoudi, Alison Brysk, and Kate Seaman, 66–92. Edward Elgar Publishing.

Ramirez, Pablo M. 2004. *El rugir de las multitudes: La fuerza de los levantamientos indígenas en Bolivia.* Ediciones Yachaywasi.

Rajan, Kaushik Sunder. 2006. *Biocapital: The Constitution of Postgenomic Life.* Duke University Press.

Ranelletti, Astride. 1934. *Solfocarbonismo: Patologia e clinica.* Atti del XI Congresso Nazionale di Medicina del Lavoro Vol. II, 73–78.

Raparelli, Francesco. 2021. *Singolarità e istituzioni. Antropologia e politica oltre l'individuo e lo Stato.* Manifestolibri.

Read, Jason. 2003. *The Micro-Politics of Capital: Marx and the Prehistory of the Present.* SUNY Press.

Read, Jason. 2017. "Anthropocene and Anthropogenesis: Philosophical Anthropology and the Ends of Man." *The South Atlantic Quarterly* 116 (2): 257–73.

Reiss, Timothy. 2003. *Mirages of the Selfe: Patterns of Personhood in Ancient and Early Modern Europe.* Stanford University Press.

Reyes, Alvaro, and Mara Kaufman. 2015. "Sovereignty, Indigeneity, Territory: Zapatista Autonomy and the New Practices of Decolonization." In *The Anomie of the Earth: Philosophy, Politics, and Autonomy in Europe and the Americas*, edited by Federico Luisetti, John Pickles, and Wilson Kaiser, 44–68. Duke University Press.

Riofrancos, Thea. 2020. *Resource Radicals: From Petro-Nationalism to Post-Extractivism in Ecuador*. Duke University Press.

Rispoli, Tania, and Miriam Tola. 2020. "Reinventing Socio-Ecological Reproduction, Designing a Feminist Logistics. Perspectives from Italy." *Feminist Studies* 46 (3): 663–73.

Roane, J. T. 2018. "Plotting the Black Commons." *Souls* 20 (3): 239–66.

Robinson, Kim S. 2017. *New York 2140*. Orbit.

Roggero, Gigi. 2011. *The Production of Living Knowledge: The Crisis of the University and the Transformation of Labor in Europe and North America*. Temple University Press.

Roggero, Gigi. 2023. *Italian Operaismo. Genealogy, History, and Method*. MIT Press.

Rossi, Ugo. 2022. "The Existential Threat of Urban Social Extractivism. Urban Revival and the Extinction Crisis in the European South." *Antipode* 54 (2): 892–913.

Rossi, Ugo, and Arturo Di Bella. 2017. "Start-Up Urbanism: New York, Rio de Janeiro and the Global Urbanization of Technology-Based Economies." *Environment and Planning A* 49 (5): 999–1018.

Rottenberg, Catherine. 2025. "The US Shifts From Progressive to Authoritarian Neoliberalism." *Counterpunch*. Retrieved from https://www.counterpunch.org/2025/02/07/the-us-shifts-from-progressive-to-authoritarian-neoliberalism/.

Saito, Kohei. 2017. *Karl Marx's Ecosocialism: Capital, Nature and the Unfinished Critique of Political Economy*. Monthly Review Press.

Salleh, Ariel. 2004. "Sustainability and Meta-Industrial Labour: Building a Synergistic Politics." *The Commoner* 9: 1–13.

Sandilands, Catriona. 1999. *The Good-Natured Feminist: Ecofeminism and the Quest for Democracy*. University of Minnesota Press.

Schabas, Margaret. 2005. *The Natural Origins of Economics*. University of Chicago Press.

Schiebinger, Londa. 1993. *Nature's Body. Gender in the Making of Modern Science*. Beacon Press.

Schnapp, Jeffrey. 1997. "The Fabric of Modern Times." *Critical Inquiry* 24 (1): 191–245.

Scott, James C. 2012. *Decoding Subaltern Politics: Ideology, Disguise, and Resistance in Agrarian Politics*. Routledge.

Sharma, Nandita Rani. 2020. *Home Rule: National Sovereignty and the Separation of Natives and Migrants*. Duke University Press.

Sharma, Nandita Rani, and Cynthia Wright. 2008. "Decolonizing Resistance, Challenging Colonial States." *Social Justice* 35 (3): 120–38.

Sharp, Hasana. 2011. *Spinoza and the Politics of Renaturalization*. University of Chicago Press.

Shiva, Vandana. 2006. *Earth Democracy. Justice, Sustainability, and Peace*. Zed Books.

Showalter, Elaine. 1993. "Hysteria, Feminisms, and Gender." In *Hysteria Beyond Freud*, edited by Sander L. Gilman, Helen King, Roy Porter, G. S. Rousseau, and Elaine Showalter, 286–44. University of California Press.

Silverblatt, Irene. 1987. *Moon, Sun, and Witches: Gender Ideologies and Class in Inca and Colonial Peru*. Princeton University Press.

Simondon, Gilbert. 2005. *L'individuation à la lumière des notions de forme et d'information*. Jerome Millon.

Simondon, Gilbert. 2009. *L'individuazione psichica e collettiva*. DeriveApprodi.

Simpson, Audra. 2014. *Mohawk Interruptus: Political Life Across the Borders of Settler States*. Duke University Press.

Singh, Neera. 2017. "Becoming a Commoner: The Commons as Sites for Affective Socio-Nature Encounters and Co-Becomings." *Theory and Politics in Organization* 17 (4): 751–76.

Slack, Paul. 2014. *The Invention of Improvement: Information and Material Progress in Seventeenth-Century England*. Oxford University Press.

Slobodian, Quinn. 2025. *Hayek's Bastards: Race, Gold, IQ, and the Capitalism of the Far Right*. Zone Books.

Smith, Neil. 1984. *Uneven Development: Nature, Capital and the Production of Space*. Blackwell.

Sotgia, Alice. 2003. "Sul filo della pazzia: Produzione e malattie del lavoro alla Viscosa di Roma negli anni Venti e Trenta." *Dimensioni e problemi della ricerca storica* 2: 195–210.

Spade, Dean. 2020. *Mutual Aid: Building Solidarity During This Crisis (and the Next)*. Verso Books.

Spillers, Hortense J. 1987. "Mama's Baby, Papa's Maybe: An American Grammar Book." *Diacritics* 17 (2): 64–81.

Spivak, Gayatri C. 1999. *A Critique of Postcolonial Reason: Toward a History of the Vanishing Present*. Harvard University Press.

Spivak, Gayatri C. 2003. *Death of a Discipline*. Columbia University Press.

Starhawk. 1988. *Dreaming the Dark: Magic, Sex, and Politics*. Beacon Press.

Stengers, Isabelle. 2005a. "The Cosmopolitical Proposal." In *Making Things Public: Atmospheres of Democracy*, edited by Bruno Latour and Peter Weibel, 994–1003. MIT Press.

Stengers, Isabelle. 2005b. "Introductory Notes on an Ecology of Practices." *Cultural Studies Review* 11 (1): 183–96.

Stengers, Isabelle. 2015. "Accepting the Reality of Gaia: A Fundamental Shift?" In *The Anthropocene and the Global Environmental Crisis*, edited by Clive Hamilton, Christophe Bonneuil, and François Gemenne, 134–44. Routledge.

Stiegler, Bernard. 1998. *Technics and Time: The Fault of Epimetheus*. Stanford University Press.

Stoler, Ann Laura. 1995. *Race and the Education of Desire: Foucault's History of Sexuality and the Colonial Order of Things*. Duke University Press.

Sultana, Farhana. 2021. "Climate Change, COVID-19, and the Co-Production of Injustices: A Feminist Reading of Overlapping Crises." *Social & Cultural Geography* 22 (4): 447–60.

Svampa, Maristella Noemí. 2019. *Neo-Extractivism in Latin America: Socio-Environmental*

Conflicts, the Territorial Turn, and New Political Narratives. Cambridge University Press.

Swyngedouw, Erik. 2006. "Metabolic Urbanization: The Making of Cyborg Cities." In *In the Nature of Cities: Urban Political Ecology and the Politics of Urban Metabolism*, edited by Nik Heynen, Maria Kaika, and Erik Swyngedouw, 36–55. Routledge.

Swyngedouw, Erik. 2011. "Depoliticized Environments: The End of Nature, Climate Change, and the Post-Political Condition." *Royal Institute of Philosophy Supplements* 69: 253–74.

Swyngedouw, Erik. 2014. "Insurgent Architects, Radical Cities and the Promise of the Political." In *The Postpolitical and Its Discontents: Spaces of Depoliticization, Specters of Radical Politics*, edited by Japhy Wilson and Erik Swyngedouw, 169–88, Edinburgh University Press.

Sze, Julie. 2018. "Vulnerable Embodiments. Denormalizing Embodied Toxicity: The Case of Kettleman City." In *Racial Ecologies*, edited by Lailani Nishime and Kim D. Hester Williams, 107–21. University of Washington Press.

Táíwò, Olúfẹ́mi O. 2024. "Climate, State and Utopia." *Boston Review*. Retrieved from https://www.bostonreview.net/forum/climate-state-and-utopia/.

TallBear, Kim. 2017. "Beyond the Life / Not Life Binary: A Feminist Indigenous Reading of Cryopreservation, Interspecies Thinking and the New Materialisms." In *Cryopolitics. Frozen Life is a Melting World*, edited by Joanna Radin and Emma Kowal, 179–202. MIT Press.

Terranova, Tiziana, and Andrea Fumagalli. 2015. "Financial Capital and the Money of the Common: The case of Commoncoin." In *Moneylab Reader: An Intervention in Digital Economy*, edited by Geert Lovink, Nathaniel Tkacz, and Patricia de Vries, 150–57. Institute of Network Culture.

Thirsk, Joan. 1964. "The Common Fields." *Past & Present* 29 (1): 3–25.

Thirsk, Joan. 1984. *The Rural Economy of England*. Bloomsbury Academic.

Thomas, Yan. 2002. Il valore delle cose. *Quodlibet*.

Thompson, Charis, and Sherilyn MacGregor. 2017. "The Death of Nature. Foundations of Ecological Feminist Thought." In *Routledge Handbook of Gender and Environment*, edited by Sherilyn MacGregor, 43–53. Routledge.

Thompson, E. P. 1993. *Custom in Common: Studies in Traditional Popular Culture*. The New Press.

Ticktin, Miriam. 2019. "From the Human to the Planetary: Speculative Futures of Care." *Medicine Anthropology Theory* 6 (3): 133–60.

Ticktin, Miriam. 2021. "Building a Feminist Commons in the Time of COVID-19." *Signs: Journal of Women in Culture and Society* 47 (1): 37–46.

Todd, Zoe. 2015. "Indigenizing the Anthropocene." In *Art in the Anthropocene: Encounters Among Aesthetics, Politics, Environments, and Epistemologies*, edited by Heather Davis and Etienne Turpin, 241–54. Open Humanities Press.

Tola, Miriam. 2016. "Composing with Gaia: Isabelle Stengers and the Feminist Politics of the Earth." *PhaenEx* 11 (1): 1–21.

Tola, Miriam. 2017. "Species, Nature, and the Politics of the Common: From Virno to Simondon." *The South Atlantic Quarterly* 116 (2): 237–55.

Tola, Miriam. 2018. "Between Pachamama and Mother Earth: Gender, Political Ontology, and the Rights of Nature in Contemporary Bolivia." *Feminist Review* 118 (1): 25–40.

Tola, Miriam. 2024. "Transcorporeità ed ecomedia in Italia: note sui materiali audiovisivi della SNIA Viscosa (1938–2015)." In *Narrazioni dall'Antropocene*, edited by Giulia Fabbri, 156–78. EditPress.

Tola, Miriam, and Ugo Rossi. 2019. "The Common." In *Keywords in Radical Geography: Antipode at 50*, edited by The Antipode Editorial Collettive, 259–63. Wiley.

Tommolillo, Niso. 2020. *Gli acidi mi hanno fatto male. Narrazioni operaie dalla Viscosa di Roma*. Il Galeone.

Tooze, Adam. 2020. "Is the Coronavirus Crash Worse than the 2008 Financial Crisis?" *Foreign Policy* (March 18, 2020).

Toscano, Alberto. 2007a. "Always Already Only Now: Negri and the Biopolitical." In *The Philosophy of Antonio Negri, Vol. 2: Lessons on Constitutive Power*, edited by Timothy Murphy and Abdul-Karim Mustapha, 109–28. Pluto.

Toscano, Alberto. 2007b. "The Disparate: Ontology and Politics in Simondon." Paper presented at the Society for European Philosophy, Forum for European Philosophy annual conference. University of Sussex.

Toscano, Alberto. 2023. *Late Fascism. Race, Capitalism and the Politics of Crisis*. Verso.

Tronto, Joan. 1993. *Moral Boundaries: A Political Argument for an Ethic of Care*. Routledge.

Tsing, Anna Lowenhaupt. 2015. *The Mushroom at the End of the World: On the Possibility of Life in Capitalist Ruins*. Princeton University Press.

Tuck, Richard. 1979. *Natural Rights Theories: Their Origin and Development*. Cambridge University Press.

Tuck, Eve, and K. Wanye Yang. 2012. "Decolonization is Not a Metaphor." *Decolonization: Indigeneity, Education & Society* 1 (1): 1–40.

Tully, James. 1993. *An Approach to Political Philosophy: Locke in Contexts*. Cambridge University Press.

Valencia, Sayak. 2018. *Gore Capitalism*. Semiotext(e). MIT Press.

Varvarousis, Angelos. 2022. *Liminal Commons: Modern Rituals of Transition in Greece*. Bloomsbury.

Vercellone, Carlo, Francesco Brancaccio, Alfonso Giuliani, and Pierluigi Vattimo. 2017. *Il comune come modo di produzione. Per una critica dell'economia politica dei beni comuni*. Ombre Corte.

Vergès, Françoise. 2017. "Racial Capitalocene." In *Futures of Black Radicalism*, edited by Gaye Theresa Johnson and Alex Lubin, 72–82. Verso Books.

Virno, Paolo. 2004. *A Grammar of the Multitude: For an Analysis of Contemporary Forms of Life*. Semiotext(e).

Virno, Paolo. 2005. Interview with Paolo Virno. Conducted by Brandon W. Joseph. Translated by A. Ricciardi. *Grey Room* 21: 26–37.

Virno, Paolo. 2008. *Multitude: Between Innovation and Negation*. Semiotext(e).

Virno, Paolo. 2009. *Angels and the General Intellect: Individuation in Duns Scotus and Gilbert Simondon*. Semiotext(e).
Virno, Paolo. 2010. *E così via all'infinito: Logica e antropologia*. Bollati Boringhieri.
Virno, Paolo. 2015. *When the Word Becomes Flesh: Language and Human Nature*. Semiotext(e).
Virno, Paolo, and Michael Hardt. 1996. *Radical Thought in Italy: A Potential Politics*. University of Minnesota Press.
Von Uexküll, Jakob Johann. 2013. *A Foray into the World of Animals and Humans*. Minnesota University Press.
Vora, Kalindi. 2015. *Life Support: Biocapital and the New History of Outsourced Labour*. University of Minnesota Press.
Wallace, Rob, Alex Liebman, Luis Fernando Chaves, and Rodrick Wallace. 2020. "COVID-19 and Circuits of Capital." *Monthly Review* 72 (1): 1–13.
Walsh, Catherine. 2016. "On Gender and Its 'Otherwise.'" In *The Palgrave Handbook of Gender and Development*, edited by Wendy Harcourt, 1–20. Palgrave MacMillan.
Webber, Jeffrey. 2011. *From Rebellion to Reform in Bolivia: Class Struggle, Indigenous Liberation, and the Politics of Evo Morales*. Haymarket Books.
Weeks, Kathi. 2011. *The Problem with Work*. Duke University Press.
Weheliye, Alexander G. 2014. *Habeas Viscus: Racializing Assemblages, Biopolitics, and Black Feminist Theories of the Human*. Duke University Press.
White, Lynn. 2003. "The Historical Roots of Our Ecological Crisis." In *This Sacred Earth: Religion, Nature, Environment*, edited by Roger S. Gottlieb, 192–201. Routledge.
Whyte, Kyle. 2016. "Indigenous Experience, Environmental Justice and Settler Colonialism." In *Nature and Experience: Phenomenology and the Environment*, edited by Bryan E. Bannon, 157–74. Rowman & Littlefield.
Wiegman, Robyn. 2000. "Feminism's Apocalyptic Futures." *New Literary History* 31 (4): 805–25.
Wilkinson, Darryl. 2013. "Politics, Infrastructure and Nonhuman Subjects: The Inka Occupation of the Amaybamba Cloud Forests." PhD dissertation. Dept. of Anthropology, Columbia University.
Winstanley, Gerrard. 2009. *The Law of Freedom in a Platform*. Oxford University Press.
Wolfe, Patrick. 2006. "Settler Colonialism and the Elimination of the Native." *Journal of Genocide Research* 8 (4): 387–409.
Wolin, Sheldon. 2004. *Politics and Vision: Continuity and Innovation in Western Political Thought*. Princeton University Press.
Wood, Ellen Meiksins. 2002. *The Origin of Capitalism: A Longer View*. Verso Books.
Wood, Ellen Meiksins. 2011. *Citizens to Lords: A Social History of Western Political Thought from Antiquity to the Late Middle Ages*. Verso Books.
Wood, Neal. 1984. *John Locke and Agrarian Capitalism*. University of California Press.
Woodly, Deva, Rachel H. Brown, Mara Marin et al. 2021. "The Politics of Care." *Contemporary Political Theory* 20 (4): 890–925.
Wright, Steve. 2002. *Storming Heaven: Class Composition and Struggle in Italian Autonomist Marxism*. Pluto Press.

Wynter, Sylvia. 2003. "Unsettling the Coloniality of Being / Power / Truth / Freedom: Towards the Human After Man, Its Overrepresentation—an Argument." *CR: The New Centennial Review* 3 (3): 257–337.

Yusoff, Kathryn. 2019. *A Billion Black Anthropocenes or None*. University of Minnesota Press.

Zechner, Manuela. 2021. *Commoning Care and Collective Power: Childcare Commons and the Micropolitics of Municipalism in Barcelona*. Transversal Texts.

Zibechi, Raúl. 2024. *Constructing Worlds Otherwise. Societies in Movement and Anticolonial Paths in Latin America*. AK Press.

Zylinska, Joanna. 2018. *The End of Man: A Feminist Counterapocalypse*. University of Minnesota Press.

INDEX

activism, 2, 3, 22–23, 73, 92, 126, 134, 141, 156, 177n6; feminist, 93–95, 138; queer, 97, 106, 108–9. *See also* B-side Pride; Ni Una Menos; Non Una di Meno
Agamben, Giorgio, 28, 34–36, 171n5, 180n5
Agard-Jones Vanessa, 119
agency: geological, 6, 79; material, 5; nonhuman, 3, 8, 29, 119; political, 8
agriculture: commercial 54, 57, 59; as improvement 51–52; industrial 40; industrialization of, 64; settler, 18; Zionist, 172n4
Alaimo, Stacy, 119, 134, 177n2
Americas, 3, 22, 37, 58, 60, 62, 67, 174n1. *See also* Indigenous
Anderson, Virginia DeJohn, 60
Andes, 152–55; precolonial, 139. See also *ayllu*
animal, 23, 27, 35, 43, 60, 65–66, 82, 103, 149, 154, 156, 173n7; animality, 79, husbandry, 41; Marx on, 64; Virno on, 77. *See also* nonhumans; species
animism, 36, 43
Anthropocene, 6–9, 22, 71–73, 76–81, 84, 91, 95–96; Anthropocene Working Group, 169n3; good, 6–8; white (M) Anthropocene, 7

anthropocentrism, 36
anthropogenesis, 78–79, 83
anthropos, 8, 71–73, 77, 80–82, 84
anti-fascism, 127–28, 178n19
appropriation, 6, 10, 14–15, 29–30, 35, 55–56, 58, 62–63, 80, 82, 98–99, 112, 145; capitalist, 11, 15, 35, 43, 79, 105, 107; colonial, 18, 45, 54, 57, 60, 67, 143; of land, 15, 17, 35, 54, 56; of nature, 98, 120, 176n9. *See also* colonialism; dispossession; enclosures; improvement; property
archive: Ex-SNIA, 125–29, 133; medieval, 20–21; reading across, 28, 44. *See also* Ex-SNIA Viscosa
Argentina, 94, 170n8, 175n6
Armiero, Marco, 131, 177n6
Arneil, Barbara, 36, 172n5
Aquinas, Thomas, 33
artificial silk, 117, 123; artificial fibers, 122. *See also* rayon
Assalti Frontali, 134
assemblage, 6, 14, 21, 27, 86, 98, 160; of dispossession, 61
assembly: activist, 118, 126; feminist, 95, 107, 165; peasant, 38–40. *See also* Forum Parco delle Energie
attachment, 74, 81, 101, 134, 136

Augustine, 31
autonomist Marxism, 2, 4–5, 13–14, 22, 72–77, 84, 90–91, 96, 102–4, 165, 169n6, 173n6, 176n11, 179n7. See also *operaismo*
autonomy: bodily, 10; conflictual, 160; from gender violence, 94; human, 79; Indigenous, 143, 145, 148; labor, 73, 75, 84, 91, 96, 104; water, 154; Zapatista, 155
ayllu, 152–54

Baccante, Maria, 127–28; Documentation Center, 118, 167
Bacon, Francis, 50–52, 172n3
Ballard, James Graham, 117
Barca, Stefania, 7, 105, 113, 119, 165
Bechtel, 151
becoming-with, 82, 90
Beguines, 32. See also heresy
Bennett, Jane, 28–29
Berardi, Franco, 91, 178n15
Berlant, Lauren, 2, 175n1
Bhandar, Brenna, 46, 52, 172n4
biopolitics, 174n11, 174n12; fascist, 121, 124; and living labor, 85–91
biopower, 85–86, 174n11
Birrell, Jean, 39
Black Mediterranean Collective, 176n16
Black Italians, 110
Blackness, 6, 79
Blanc, Paul David, 121, 123
Blaser, Mario, 18, 140, 144–45, 147, 156, 163, 179n6
Blith, Walter, 49–50
Bodin, Jean, 43
body-territory, 108, 176n14
Boisseron, Bénédicte, 173n7
Bolivia, 23, 140–41, 145, 151–55; constitution, 151. See also Cochabamba; water committees, water-war
Bonaventura, 33–34
Bonneuil, Christophe, 7

Brace, Laura, 50, 55
Bratich, Jack, 166, 180n2
Braun, Bruce, 75, 135, 165
Briggs, Laura, 177n3
Bullicante Lake, 130, 134–35, 138, 160–61. See also Ex-SNIA Viscosa; natural monument; nonhumans; Rome
Burr, David, 34, 171n4
Butler, Judith, 89
Bynum, Caroline, 15, 29, 42, 44, 171n10
Byrd, Jodi, 18–19, 141–42
B-side Pride, 108–9

Caffentzis, George, 147. See also Midnight Notes Collective
Campi Aperti, 109
Campidoglio, 161
Capital (Marx), 12, 39, 62, 64
capitalism: agrarian, 56; cognitive, 72, 173n2; colonial, 44–45, 73, 107, 144; critique of, 14, 21, 74–75, 92, 174n15; disaster, 12; extractive, 14, 90; fossil, 171n4; industrial, 15–16, 56, 62–63, 76, 87; and nature, 88; neoliberal, 12, 107, 135, 139; post-Fordism, 74, 76–78; racial, 176n16. See also extractivism
capitalist accumulation, 8, 14, 28, 47, 59, 74, 76, 80, 82, 85, 88, 92. See also primitive accumulation
Capitalocene, 105; racial, 7
caracoles, 146; Oventic, 149
carbon disulfide, 121, 123–25. See also chemicals; SNIA Viscosa; toxic embodiment; toxicity
care: and Black feminism, 98; crisis, 95, 108; debt, 112–13; earthcare, 103; and environments, 105; ethics of, 175n7; global care chains, 99, 102; inequalities, 95, 106; infrastructures, 12, 21, 103; radical, 107, 112; trans, 109; transfeminist, 111; work, 14–15, 95, 99, 104, 111, 175n7. See also reproduction

Care Collective, 95, 107
Casalini, Brunella, 175n7
Casarino, Cesare, 88–89
Cassata, Francesco, 124
Cathars, 32. *See also* heresy
Celano, Thomas, 35. *See also* Francis
Cellamare, Carlo, 3
Césaire, Aimé, 67
Chakrabarty, Dipesh, 78
Charbonnier, Pierre, 172n1
Charter of the Forest, 38–39, 142
chemicals: embodiment, 119–20, 123, 125, 135; exposure, 117, 124; infrastructure, 118–20, 125–26, 128–29; poisoning, 121, 124, 128–29; sensations, 125, violence, 120, 125. *See also* SNIA Viscosa; toxic embodiment; toxicity
Chen, Mel, 119
Chiapas (Mexico), 101, 137–38, 141, 145–46, 150, 155. *See also* Zapatistas
Chomsky, Noam, 142
Cirillo, Valeria, 178n24
class, 47, 83, 95, 98, 104; antagonism, 121; gender, race and, 3, 8, 106, 110–11, 119, 140, 163; middle, 109; ruling, 59; struggle, 21, 153–54, 170n7; working, 5, 12, 59, 75, 88
climate: change, 1–2, 7, 78, 90, 105, 107; crises, 160; disaster, 155; injustice, 24, 159; justice, 21, 138; policy, 180n1
Cochabamba, 145, 151–52, 154. *See also* Bolivia
Cohen, Ed, 45, 66, 156, 166
Cohen, Jeffrey Jerome, 171n10
Colectivo Miradas Criticas del Territorio desde el Feminismo, 108
Collard, Rosemary-Claire, 96, 104
colonialism, 102, 172n6, 176n16; encounters, 45; European, 8–9, 16, 36–37, 61–62, 100, 153, 156–57; Franciscan, 37; in Ireland, 47, 51; patriarchy and, 21; settler, 17, 119, 142. *See also* habitation; modernity; race and racism

Combes, Muriel, 83, 174n10
commons, 1–16, 18–24, 27–30, 37–44, 46–50, 54, 57–62, 66–68, 72–73, 75, 84, 90, 93–98, 101–7, 109–14, 118–19, 131–32, 134–36, 138–47, 150–51, 153–56, 159–63, 165; colonial, 59–60; common, 13–14, 73; commoners, 27, 37–40, 44, 57–58, 62, 67, 163; commoning, 2–3, 6, 9, 12–16, 18–19, 23–24, 27–28, 40, 44, 59, 61, 63, 92, 94–97, 103–4, 106, 109, 111, 113–14, 117–18, 120, 131–32, 135–36, 141, 145, 157, 160, 163; common-pool-resources (CPR), 4, 11, 16; common use, 8, 15, 19–22, 28–31, 33, 35–36, 42–45, 60, 141, 144; cosmopolitical, 118, 135; ecology of, 13, 73, 84; institutions, 11, 113, 160; latent, 162; medieval, 16, 21, 27–29, 38–38, 44; more-than-human, 8, 27; natural, 22, 87; paradoxes of, 9, 18, 49; resurgent, 44, 160–63; social 22, 87; socio-ecological, 118; transfeminist, 3, 23, 93, 95–96, 110–14; uncommons, 144–45, 156; and universalism, 140–41, 145, 157; urban, 2, 94, 132, 169n5
Common Coin, 131
Communist Manifesto (Marx and Engels), 63
community: Christian, 31–32; earth, 15, 101; economies, 175n1; Indigenous, 153; medieval peasant, 10, 30, 36, 42–43; political, 17, 80, 150; Zapatista, 134, 149, 150
consensual invasion, 23, 139
Conte, Alessandra, 126, 177n7, 178n11, 178n12
Conventuals, 33–34. *See also* Franciscans
Cooper, Melinda, 4, 105
Coordinadora de Defensa del Agua y de la Vida (Coalition for the Defense of Water and Life), 152, 154. *See also* Bolivia; water committees, water-war

cosmology: Andean, 152–54; European peasant, 41; Indigenous, 147, 149–50, 152, 154. *See also* political ontology
cosmopolitics, 118–19, 133–35
Coulthard, Glen, 6, 18, 61, 143, 148, 156
COVID-19, 93, 95, 97, 106–8. *See also* zoonosis
custom, 36–40, 42, 44

Dakota Access Pipeline, 140, 143
Dalla Costa, Mariarosa, 14, 23, 98–99, 101
Daston, Lorraine, 41
Davenant, Charles, 46–47
debt: care, 112–13; public, 94, 117; owed by capitalism, 100
decolonial perspectives, 3, 5, 9, 23, 60–61, 139, 141, 143–45, 176n14, 179n3, 179n7, 180n1; decolonial reflexivity, 142
decommodification, 104, 136; urban, 94
DeJohn Anderson, Virginia, 60
de la Cadena, Marisol, 18, 140, 144, 147, 154, 156
Deleuze, Gilles, 71, 85–86, 161
Dempsey, Jessica, 5, 96, 104
denaturalization, 90, 174n13
De Rosa, Salvatore Paolo, 177n6
Di Chiro, Giovanna, 6–7, 71
Diggers, 50
disaster: capitalism, 1; environmental, 107, 155; management, 105
discipline, 20; capitalist, 74, 76, 117
dispossession: capitalist, 12–13, 59; colonial, 22, 49, 61, 143; ecological, 7, 104
Dominicans, 32–34
Dowling, Emma, 95
Duns Scotus, 33

earth: as alterity, 104; compared with global, planetary, 104; feminized, 50, 64, 104; habitable, 95, 106, 113; productivity, 50, 53, 58

Earth Systems, 71
ecofeminism, 98–101, 98. *See also* feminist theory; reproduction
ecology of dispossession, 47–48, 53, 58, 60, 62, 66–68. *See also* improvement
eco-memory, 126, 129
economy: classic political, 10, 17, 40, 65; extractive, 3–4, 145–46; monetary, 39, 57, 144; self-sufficiency, 122; of subsistence, 3, 15, 36–40, 54, 59, 67, 100–3
Ecuadorian Amazon, 140
Eidelman, Tessa, 169n5
ejidos, 146
Empire (Hardt and Negri), 84–85, 91
enclosures, 3, 8, 9–10, 12, 17–18, 22, 28–29, 35–37, 40, 43, 45, 47–52, 98, 159; Locke and, 54–62; new, 156; ongoing 28, 63, 101, 146, 150, 156, 169n5
environmental humanities, 6, 71, 165, 169n4; feminist, 119, 163, 169n4
environmental violence, 3, 100, 103, 114
environmentalism, 36; of the poor, 101; worker, 75. *See also* climate
Esteva, Gustavo, 147
Esposito, Roberto, 170n3, 180n7
Ethiopia, 122
European modernity, 4, 8–9, 19–20, 27, 42, 44–45, 62, 71, 144, 146. *See also* colonialism
European South, 2, 23, 95, 114
exploitation: gendered, racialized, 5, 14–15, 54, 59, 63, 97–99, 100–2, 104, 108; labor, 17, 23, 66, 75, 79, 88; of nature, 100, 103, 108
expropriation: of the commons, 12, 16, 58, 90, 101
Ex-SNIA Viscosa, 23, 117–19, 125–26, 128–30, 132–37, 177n4, 117n8. *See also* archive; Bullicante Lake; commons; Rome; SNIA Viscosa
extractivism, 90, 140; expanded, 95; urban social, 2; zone, 180n1

208 *Index*

Farris, Sara, 180n4
fascism, 2, 21; governance, 119–25; modernity 122–23. *See also* antifascism; biopolitics
Federici, Silvia, 9, 12, 14–15, 28–29, 43, 98–103, 170n8, 171n11, 174n1, 176n11. *See also* re-enchantment; reproduction; Wages for Housework
femicide, 95. *See also* gender: violence
feminist international, 94
feminist strike, 107
feminist theory: ecofeminism, 98–101; Latin America, 97; Marxism feminism, 14
femonationalism, 180n4
Ferdinand, Malcolm, 7, 54, 72, 177n18
Fordism, 74, 76, 128, 173n4; post-, 74, 76–79, 82, 87, 178n15
form of life, 78; Franciscan, 32, 35, 37
Fortier, Craig, 170n10
Forum Parco delle Energie, 125–26, 130, 132, 178n10
fossil fuel, 1, 75, 90, 160
Foucault, Michel, 20, 28, 32, 66, 85–86, 89, 125, 129, 174n11, 174n12
Fox Keller, Evelyn, 172n6
Fracassi, Matilde, 125–26, 128, 131, 177n5, 178n14, 178n21
"Fragment on Machines" (Marx), 75–76. See also *Grundrisse*
Fragnito, Maddalena, 95, 111, 165
Francis, 32, garden, 35–36
Franciscans, 21, 32–37, 144. *See also* poverty
Fraser, Nancy, 95–98

Gaard, Greta, 176n9
Gago, Verónica, 4, 90, 94–95, 105, 140, 170n8
Garnsey, Peter, 31
Gehlen, Arnold, 77. *See also* philosophical anthropology
gender, 89, 176n9; binarism, 175n4; and coloniality, 137, 179n1; dissidents, 21, 93; and improvement, 48, 51, 52–53; and labor, 14–15, 17, 29, 95, 98–99, 102, 104, 111, 121; and nature, 43–44, 50, 172n3; and property, 22; race, and species, 78–79; violence, 94–96, 106, 110, 112. *See also* care; reproduction
genealogy (as method), 20, 28, 174n12
general intellect, 75–76, 91
Gibson-Graham, J. K., 8, 118
Ginzburg, Carlo, 15, 29, 41, 171n9
Gómez-Barris, 180n1
governance: fascist, 119–20; from below, 118, 131; global, 85; neoliberal, 132–33; Ostrom on, 11–12, 72; self, 2–3, 13, 161; state and market, 4; Zapatista, 143, 147–48, 155
Graeber, David, 177n17
Greer, Allan, 18, 58–60, 142
Grossi, Paolo, 171n5
Grosz, Elizabeth, 165, 174n10
Grundrisse 66, 75
Gutiérrez Aguilar, Raquel, 174n1

habitation, 112; colonial, 54, 177n18; urban, 113. *See also* inhabiting
Hamilton, Jennifer, 169n4
Haraway, Donna, 8, 65, 71, 84, 90, 177n1
Hardin, Garrett 2, 9–11, 28
Hardt, Michael, 4, 77, 166; and Negri, 9, 22, 72–73, 84–91, 102, 118, 132, 151
Hartlib, Samuel, 51–52
Harney, Stefano, 67
Harvey, David, 4, 12, 28, 118, 132–33, 173n4
health: care, 75, 95, 106, 107–8, 112; and environment, 75; public, 74; risks in the workplace, 124, 127, 177n8
heresy, 28, 32, 36, 144; poverty as, 30, 33–35. *See also* Beguines; Cathars; Dominicans; Franciscans; Waldensians
Hobbes, Thomas, 17, 80
Hoffman, Richard, 29, 36, 41

Index 209

homo oeconomicus, 10
homo sapiens, 6, 22, 65–66, 73, 77–79, 82, 84
human, humanity: dominant model of, 5–6, 8–9, 17–18, 22, 48, 63, 67, 173n7; and nature, 41–42, 47, 82, 85; as species, 19, 65–66, 71, 77–78, 79–80
human exceptionalism, 22, 73, 91, 96
humanism, 89; exclusionary, 144; liberal, 17
Huron, Amanda, 4, 9
housewifery, 51
husbandry, 50–51
hysteria, 123, 177n3

imperium, 30
improvement: land, 17, 22, 46–53, 61; Locke on, 47, 54, 56–58, 61; Marx on, 61–63
Inca empire, 139
Ince, Onur Ulas, 58, 172n6
inhabiting, 9, 23–24, 27, 35, 41, 45, 61, 67, 73, 113–14, 161. *See also* habitation
Indigenous: eco-memories, 129; land, land relations, 12, 18–19, 22, 45, 47–48, 54, 56–60, 62, 67, 140, 142, 144, 146, 148, 153–54, 170n10; sovereignty, 19, 141, 143, 145, 179n4; Standing Rock, 140, 143; struggles, 19, 23, 85, 139–143, 147, 152, 156; studies, 18, 139; women, 108, 138. *See also* autonomy; Bolivia; Chiapas; colonialism; cosmology; political ontology; Zapatistas
Indigenous National Congress (Mexico), 138
individuation, 73, 80–83. *See also* preindividual; Simondon; transindividual
Inquisition, 41, 171n9
institutions, 114, 131, 163; of the commons, 12, 113, 160–61; feminist, 95
intersectionality, 110, 112
interspecies: dependencies, 109, 129;

hierarchies, 91. *See also* multispecies; species
Italy, 2, 107–8, 110, 175n4; 1970s struggles, 74; far-right, 21, 160; feminist activism, 94–96; rayon industry, 122; resurgent commons, 131

Jackson, Steven, 97
Jackson, Zakiyya Iman, 170n2, 173n7

Kaika, Maria, 2
Kauanui, J. Kēhaulani, 18, 142, 170n9
Kennedy, Rosanne, 129
Kiser, Lisa, 35, 41
Kohn, Margaret, 147

labor: gendered division of, 51; hybrid, 105; living, 4, 13–16, 63, 65, 75–76, 84–87, 89, 91, 105, 180n7; Locke on, 54–57; precarity, 23, 118, 131, 135; reproductive, 14–15, 43, 53, 98–100, 104, 111; and toxicity, 23, 117–18, 123, 128
Lancione, Michele, 161, 177n19, 180n6
land: access, 3, 16, 28, 37, 44, 50, 58, 103, 104, 143; Indigenous, 12, 18, 22, 45, 48, 54, 57–61, 67, 140, 142, 144; vacant, 57; Zapatista politics of, 141, 147, 149–51, 156; *See also* improvement; Indigenous
Laslett, Peter, 56
Latin America, 90, 108, 179n7, 180n1; feminist movements in, 94, 97, 102; Indigenous politics, 12, 143–44, 147
Leff, Gordon, 34
LeMenager, Stephanie, 169n1
Leonardi, Emanuele, 75, 91–92, 165, 174n14
Liboiron, Max, 179n3
Linebaugh, Peter, 12, 27, 39–40, 49, 59. *See also* Midnight Notes Collective
Linnaeus, Carl, 65
Livingston, Julie, 27, 98

Locke, John, 17, 22, 45, 47–48, 53–58, 80, 88, 172n5; *See also* improvement; labor, property
Lorde, Audre, 98
Loriga, Giovanni, 123
Lowe, Lisa, 17, 20, 28, 98
Lucha y Siesta, 93–97, 107, 110–13, 160, 165–6, 175n2, 176n15. *See also* activism; commons: transfeminist; Rome
Lugones, Maria, 179n1

Macao, 131
magic, 28–29, 41, 43–44, 126
Magna Carta, 38–39, 142
Major Ana Maria, 146, 149, 150
Malatino, Hil, 14, 109, 176n8
Mamani Ramirez, Pablo, 152–53
Manjapra, Kris, 54
many-headed hydra, 59
Marcos, Subcomandante, 147–49, 179n9
Marcos, Sylvia, 137, 179n8
Marichuy (María de Jesús Patricio Martínez), 138
Marinetti, Filippo Tommaso, 123
Markham, Gervase, 51, 172n2
Marx, Karl, 12–14, 22, 28, 39–40, 49, 52, 62–68, 75, 78, 84, 86, 88–89, 99, 102, 170n7, 172n7, 173n8. *See also* autonomist Marxism
materialism, 29, 41; historical, 78; new, 170n2
materiality, 88, 113, 170n2; medieval, 16, 21, 171n10
matter, 28–29, 42, 43–44, 64, 79, 100, 140, 171n10, 176n10
Mbembe, Achilles, 79
McBride, Kelly, 147
McClure, Julia, 37
McCormick, Ted, 53
Mehta, Uday, 48
Meloni, Giorgia, 21, 160, 180n4
memory, 44, 117–28, 133, 135; of combats, 28, 161; embodied, 42

Mentinis, Mihalis, 149
Merchant, Carolyn, 99, 101, 172n3
metabolic rift, 63, 172n7
metabolism, 64–65, 88, 178n22; metabolic flows, 100, 118; metabolic relation, 131, 133
meta-industrial work, 100
Mexico, 19, 23, 102, 137–39, 141, 145–46, 148
Meziane, Mohamed Amed, 171n8
Mezzadra, Sandro, 4, 13, 90, 105, 135, 140
microfascism, 160, 180n2
Middle Ages, 42, 44, 171n10
Midnight Notes Collective, 12, 102, 146–47
Mies, Maria, 12, 14–15, 43, 98–101, 103, 105, 171n11
Mignolo, Walter, 143, 147, 149–50
Militant A, 134
modernity, 6–7, 9, 61, 80, 89, 91, 100, 122, 140, 148; European, 4, 8–9, 19–20, 27, 42, 44, 45, 53, 62, 71, 144, 146; liberal, 16; Western, 2, 19, 76, 144, 148, 179n7
Molavi, Shourideh, 180n3
Moore, Jason, 15, 79, 105, 120
monasticism, 31–32
Montaña, La, 137
Morales, Evo, 145, 152
more-than-human: commons, 23, 27; world, 28, 37, 44, 97. *See also* nonhumans
Moten, Fred, 67
Moulier Boutang, Yann, 173n2
multispecies, 60, 174n13
multitude, 80, 82, 85, 88
Muñoz, José Esteban, 175n1
Murphy, Michelle, 119, 120, 126
Muro del Canto, 134
Mussolini, Benito, 120–24
mutual-aid, 93, 97, 106
Münzer, Thomas, 40

natural monument, 131, 178n20
natural right, 22, 30, 45, 47, 54–55

nature, 4, 6–10, 13–14, 17, 19, 22, 29, 30, 36, 40–41; as assemblage, 6; cheap, 7, 105; Hardt and Negri on, 85–91; improvement of, 50–58, Marx on; 22, 49, 63–66, 78; mastery of, 36, 99, 160; materialist ecofeminism on, 90, 101–5; medieval, 29, 41, 43–44; metabolic relation to, 132–33; Simondon on, 80–83; second, 134; third, 135; urban, 3, 5, 133, 135; Virno on, 72, 76, 78–80, 82, 84. *See also* improvement; land; materiality; more-than-human; nonhumans

Negri, Antonio, 4–5, 9, 13, 22, 72–76, 84–91, 102, 118, 132, 151, 161, 173n2, 173n3, 180n7

Neimanis, Astrida, 7, 168n4

Nelson, Sara, 75, 165, 176n11

neoliberalism: and authoritarianism 3, 15, 180n2; late, 112

Neyrat, Frédréric, 7, 72, 104

New York 2140 (novel), 1–3, 159

Ni Una Menos, 94. *See also* activism; Latin America

Nixon, Rob, 10, 120

nonhumans: actors, 20, 23, 27, 29, 40, 43, 108, 126, 136, 139, 141; agency, 3, 119; humans and, 6, 8–9, 13, 18, 48, 53, 65, 83, 96, 98, 100, 103, 112, 135, 145–46; productivity, 67; work, 104–5. *See also* more-than-human; nature

Non Una di Meno, 96, 107–8, 110, 113, 165. *See also* activism; Latin America; Ni Una Menos

occupational medicine, 123–24

Olivera, Marcela, 152, 154

Olivi, Peter, 34

omnia sunt communia, 40

ontogenesis, 81

onto-epistemic worlds, 139

ontology of the present, 20. *See also* genealogy (as method)

operaismo, 74–75, 169n6

Osco, Marcelo Fernandez, 153

Ostrom, Elinor, 4–5, 9, 11–12, 27, 72

otroas, 137–38

Pachamama, 152–54

Palestine, 160, 172n4, 180n3

pandemic, 95, 97, 106–9, 175n5

Papadopoulos, Dimitris, 8, 178n18

Parco delle Energie, 118, 126

partition of the sensible, 18

partnership ethics, 101

Pasolini, Pier Paolo, 121

passibility, 16, 44

Pellizzoni, Luigi, 7, 105, 135, 165, 170n2, 173n8

perambulation, 42–43

personhood, 15–16, 18, 47, 55

Petersmann, Marie, 178n20

Petruccioli, Rita, 175n3

Petty, William, 46–47, 51–53, 64, 121, 172n4

philosophical anthropology, 77

Pinkus, Karen, 122

place-based, 144, 154, 162

placemaking, 96, 113

planetary, the 104

plantation, 4, 14, 54, 99, 138

Plessner, Helmut, 77

pluriverse, 139

political arithmetic, 52

political ecology (as perspective), 5, 118, 119, 133, 165

political ontology, 17, 79, 144, 146, 150, 155–56, 179n6

population, 174n11; growth, 9–11, 39, 43; improvement of, 46–47, 52–53

Porto Marghera, 74–75

possessive individual, 17, 58, 62, 80, 92

post-Fordism, 74, 76–79. *See also* Fordism

Posthumus, Stephanie, 112

Potere Operaio, 74, 101

poverty: dispute, 30, 32–34, 144. *See also* Dominicans; Franciscans
Povinelli, Elizabeth, 20, 140, 173n6
power of life, 85–86
precarity, 2, 108, 112, 117–18, 160, 163, 178n15; ecological, 95, 109; labor, 23, 78, 131, 135
prefiguration, 12, 23, 110, 177n17
preindividual, 81–84
premodern Europe, 10, 19, 29, 44
primitive accumulation, 12, 16, 40, 43, 62–63, 171n8
privatization, 3, 10, 18, 27, 95, 112, 161; neoliberal, 101–3, 112 141, 152–54; of social services, 103, 112; urban 135; of water and land, 3, 12, 40, 59, 102, 131, 141, 145, 147, 151. *See also* appropriation; enclosures
production of nature, 88–89
properly political, 118, 133
property, 3, 11, 15–18, 20, 23, 30–33, 40, 43, 50–51, 53, 61–62, 66–67, 94, 119, 130–32, 134, 142, 144, 155, 162; Franciscan renunciation of, 31, 33–38, 171n5; Locke's theory of, 22, 45, 48, 54–58; private, 12, 31, 36, 50, 54, 58–59, 61, 66, 88, 90, 96, 111, 113, 131, 153, 163; racial, 52, 172n4. *See also* appropriation; enclosures; privatization
Puig de la Bellacasa, Maria, 178n18

race and racism, 94–95, 98, 108, 110, 153; and improvement, 47–48, 53, 172n4; in Italy, 110, 176n16; racial capitalocene, 7; racial hierarchies, 3, 5, 8, 78, 83, 94, 106, 111, 120, 140, 156, 172n5; racial violence, 4; racialized accumulation, 142; racialized bodies, 5, 8–9, 95–97, 99, 102, 108, 110; racialized labor, 63, 99, 104, 171n8; and species, 65–67, 79, 84, 173n7. *See also* colonialism; plantation; racial ownership; slavery; species

Ranelletti, Aristide, 124
Raparelli, Francesco, 161, 166, 180n7
rayon, 122–23; industry, 120–22, 124–27, 129, 177n4. *See also* artificial silk; viscose
Read, Jason, 165
recognition, 147–48, 152, 161
Rediker, Markus, 48, 58–59
re-enchantment, 29, 102
Reiss, Timothy, 16, 44
repair, 21–23, 93, 96–97, 100, 111, 113–14; ecological, 130, 178n18
reproduction, 6, 14, 21, 23, 43, 146, 155; fascistization of, 170n8; forces of, 105; reproductive labor, 14–15, 98–100; social and ecological 14–15, 95–109, 113–14, 133, 154, 165, 175n7. *See also* care: work; Wages for Housework
res communis, 10, 30
resilience, 3, 113
res nullius, 10, 30
Rete Commons (Naples), 131
Robinson, Kim Stanley, 1–3, 159
Rome, 2–3, 8, 19, 23, 31, 93, 110, 117, 120–22, 124–25, 128, 131, 134, 138–39, 159–63, 165–67; as self-made city, 3
Rosenberg, Jordy, 170n2
Rossi, Ugo, 3, 11, 13, 112, 165
ruins: ancient, 148, 163; capitalist, 1, 135, 162–63; capitalist ruination, 21, 107; post-industrial, 3, 23, 118–19, 126, 134

Safransky, Sara, 169n5
Saito, Kohei, 63, 172n7
Salleh, Ariel, 14, 98–100, 105
Schiebinger, Londa, 65
Schiwy, Freya, 149–50
Schnapp, Jeffrey, 122–23
Scott, James, 38
settler colonialism, 17–18, 119, 141–42, 179n5
Sharma, Nandita, 141, 179n4
Sharp, Hasana, 84, 174n10

Index 213

Shiva, Vandana, 14–15, 98, 101
Sicuriello, Flavia, 129, 178n16
Simondon, Gilbert, 73, 80–84, 173n8, 174n9, 174n10
Simpson, Audra, 140, 142
Slack, Paul, 46–47, 49, 52
slavery, 8, 14, 37, 54, 62, 67, 100, 102, 107, 142, 171n11; slave trade, 17, 45, 48, 172n5
smell, 125–6, 177n6
Smith, Adam, 66
Smith, Neil, 88, 134
SNIA Viscosa, 119, 122–23, 125. *See also* chemicals; Ex-SNIA Viscosa; toxicity
social cooperation, 11, 13, 19, 22, 76, 85, 87, 90, 96, 118, 156; worker's, 5. *See also* labor: living
species, 3, 5, 6, 22, 86, 95, 108, 111, 136, 156; human, 71–72, 77–80, 82–84, 161; laboring, 19; speciation, 67; species-being, 64–66. *See also* interspecies; multispecies
Spirituals, 33–4. *See also* Franciscans
Spivak, Gayatri, 66, 104
Starhawk, 43–44
state of nature, 17, 49, 54, 56–57, 67, 80
Stengers, Isabelle, 7, 44, 118, 129, 133–34, 170n7, 170n1, 174n15, 176n10, 177n1
Stock, Mathis, 112
subjugated knowledges, 20. *See also* genealogy (as method)
submerged perspectives, 180n1
subsistence perspective, 103. *See also* economy: of subsistence
Swyngedouw, Erik, 7, 88–89, 118, 133
Sze, Julie, 119

Táíwò, Olúfẹ́mi O., 180n5
Teatro Valle, 131
temporalities, 19, 96, 106
thingification, 67
Thirsk, Joan, 37, 40, 46
Thomas, Yan, 170n3

Thompson, E. P., 15, 38, 42
Ticktin, Miriam, 4, 107, 114
tirakuna, 154
Tommolillo, Niso, 121, 177n4
Tooze, Adam, 95
Toscano, Alberto, 21, 83, 86, 174n12
toxicity, 7, 23, 74, 101, 117–20, 123, 125, 128, 135. *See also* Porto Marghera; SNIA Viscosa
toxic embodiment, 118–19, 121. *See also* chemicals; SNIA Viscosa
"Tragedy of the Commons, The" (Hardin), 9–10
transcorporeality, 123, 177n2
transfeminism, 175n4. *See also* activism; commons: transfeminist; Lucha y Siesta; Ni Una Menos; Non Una di Meno
transindividual, 81–82
Tronto, Joan, 98, 111–12
Tsing, Anna, 28, 107, 118, 135, 162
Tuck, Eve, 60, 179n3
Tuck, Richard, 30, 33–34
Tully, James, 56

umwelt, 77
urban natures, 3, 5, 23, 75, 129, 135
urban resettlement, 120
urban speculation (in Rome), 120, 130. *See also* Ex-SNIA Viscosa

Valencia, Sayak, 137, 175n4
Vercellone, Carlo, 4, 9–10, 12–13, 72, 96
Vergès, Françoise, 7, 71
violence: chemical, 120, 125; environmental, racial, gendered, 1, 3–5, 8, 23, 71–72, 91, 93–96, 99, 100, 103, 106, 107, 109, 111, 112, 114, 137, 155–56, 161, 163, 175n2; of primitive accumulation, 40, 43, 49, 62
Virno, Paolo, 4–5, 13, 22, 72–74, 76–80, 82–84, 91
viscose: industry, 121, 124, 129, produc-

tion, 121, 125. *See also* artificial silk; rayon
von Uexküll, Jakob, 77
vulnerabilities, 108, 118, 129

Wages for Housework, 14, 99, 101. *See also* care: work; labor: reproductive; reproduction
Waldensians, 32
Walking Forest (The), 161
Walsh, Catherine, 179n1
waste, 7, 71, 75, 85, 120, 125; as uncultivated land, 3, 17, 35, 37–38, 40, 48–50, 52, 55–58
water committees (Bolivia), 151–52, 154–55
water-war (Bolivia), 152, 154
Webber, Jeffrey, 152–53
Weeks, Kathi, 13
welfare state, 2, 106, 113, 117, 161
White femininity, 160
White man, 5, 17–18, 22, 45, 48, 66, 71
White, Lynn, 36
Whiteness, 110, 142
Whyte, Kyle, 140
Winstanley, Gerrard, 50–51
witches, 27–28, 44; witchcraft, 43; witch hunt, 43, 171n11

Wolin, Sheldon, 30–31
Woods, Neal, 56
work: essential, 106, 108; industrial 117–18, 121, 122–30; reproductive, care 14–15, 43, 51, 95, 99–100, 102–4, 111; workforce, 7, 62, 120, workers' struggles, 73–75, 86, 117, 120, 127–28. *See also* care; class; labor; reproduction
worker environmentalism, 75. *See also* Porto Marghera
workerism, 74–75, 86, 169n6, 179n7. *See also* autonomist Marxism; Potere Operaio
World Bank, 102, 151
Wynter, Sylvia, 71

Yang, Wayne, 61, 179n3
Yusoff, Kathryn, 8

Zapatistas, 23–24, 137–39, 141, 143, 145–51, 155–57; cosmology, 147, 149–50; EZLN (Ejército Zapatista de Liberación Nacional), 146; Women's Revolutionary Law, 146. *See also* Indigenous; political ontology
zoonosis, 107
Zylinska, Joanna, 6

Miriam Tola is Associate Professor in Communication and Media Studies at John Cabot University. She is coeditor of *The Routledge Handbook of Ecomedia Studies* (2023) and *Ecologie della cura: Prospettive transfemministe* (2021).

MEANING SYSTEMS

The Beginning of Heaven and Earth Has No Name: Seven Days with Second-Order Cybernetics. Edited by Albert Müller and Karl H. Müller. Translated by Elinor Rooks and Michael Kasenbacher.
HEINZ VON FOERSTER

Cultural Techniques: Grids, Filters, Doors, and Other Articulations of the Real. Translated by Geoffrey Winthrop-Young.
BERNHARD SIEGERT

Interdependence: Biology and Beyond.
KRITI SHARMA

Earth, Life, and System: Evolution and Ecology on a Gaian Planet.
BRUCE CLARKE (ED.)

Upside-Down Gods and Gregory Bateson's World of Difference.
PETER HARRIES-JONES

The Technological Introject: Friedrich Kittler between Implementation and the Incalcuable.
JEFFREY CHAMPLIN AND ANTJE PFANNKUCHEN (EDS.)

The Unconstructable Earth: An Ecology of Separation. Translated by Drew S. Burk.
FRÉDÉRIC NEYRAT

Google Me: One-Click Democracy. Translated by Michael Syrotinski.
BARBARA CASSIN

Exterranean: Extraction in the Humanist Anthropocene.
PHILLIP JOHN USHER

Expanded Cinema, Fiftieth Anniversary Edition. Introduction by R. Buckminster Fuller.
GENE YOUNGBLOOD

Cybernetic Capitalism: A Critical Theory of the Incommunicable.
JAN OVERWIJK

Resurgent Commons: Feminist Political Ecologies in the European South.
MIRIAM TOLA

www.ingramcontent.com/pod-product-compliance
Lightning Source LLC
Chambersburg PA
CBHW031148020426
42333CB00013B/562